Studying Sound

A Theory and Practice of Sound Design

Karen Collins

The MIT Press
Cambridge, Massachusetts
London, England

This book was set in ITC Stone Serif Std and ITC Stone Sans Std by New Best-set Typesetters Ltd. Printed and bound in the United States of America.

Library of Congress Cataloging-in-Publication Data is available.

ISBN: 978-0-262-04413-4

10 9 8 7 6 5 4 3 2 1

Studying Sound

Contents

Acknowledgments

This book would not have been possible without the many students who contributed to my lectures in sound design at the University of Waterloo and guest lectures around the world. I am grateful to their input on learning methods and the discussions we had about sound design. I also must acknowledge the input of those who contributed to lectures and workshops at academic and industry conferences and organizations where the teaching of sound was often at the forefront.

Some of the ideas presented here are based on work I have produced with coauthors, colleagues, and supervisors over the years, and I must also acknowledge their intellectual contributions to my research in sound: Bill Kapralos, Philip Tagg, Paul Théberge, Ruth Dockwray, Alexander Hodge, Michael Dixon, and Kevin Harrigan: thank you!

The accompanying photos and videos found on the studyingsound.org website were taken with Andrew Smith and Charlotte Baker, who also took the photos for this book. They also contributed most of the audio examples online, and Charlotte built out the website.

The funding for some of the experiments I have undertaken over the years that contributed to my thinking about sound was provided primarily by the Social Sciences and Humanities Research Council of Canada. I am very grateful for their continued support.

Introduction

The composer John Cage tells the following story in his book *Silence*: after he played the same sound on a loop nonstop for fifteen minutes to a class of students, a woman got up and ran screaming from the room, "Take it off, I can't bear it any longer!" Cage turned the sound off, only for another student to ask, "Why'd you take it off? It was just starting to get interesting" (2013, 93). Throughout this book, I hope that you will learn to "find the interesting" in sound. I aim to take you on a journey from being the person who might run out of the room screaming in annoyance, to being someone who is very comfortable thinking and talking about sound, who can focus on sound and learn to listen to different aspects of sound. A person who finds that the more they listen, the more interesting their world becomes.

Other books about sound design are available, but in my experience, sound design tends to be taught as sound for moving image (that is, sound for film/television, animation, theater, or games). The reader is left with no time to cultivate an appreciation of just *sound*, or to develop a language and rhetoric of sound on its own, to explore the potential that lies in sound as a medium and as a rhetorical device. The complexity of sound on its own is often rushed through in order to get to the technical aspects of sound for moving image. Part of this oversight is the fault of an educational system that focuses on other aspects of the production of media, and the entire world of sound is often forced into a thirty-hour course over one semester. The result is that sound designers are always enslaved to the image, creating sound for that purpose, rather than developing their skills in actually designing sound. I'm not suggesting that sound design for moving image doesn't have a purpose: clearly, it has a very specific purpose. I've spent years studying the relationship that sound has to image, and image to sound, and at this point published seven books and countless research papers on that very subject. However, sound for moving image has often assumed the role of a subordinate: sound is there, we are told, to support the dominant image. The eye rules supreme in our ocular-centric Western culture. Is it any surprise that image dominates sound design practice and education, too? Of course, most sound design jobs are in film or games, so it's understandable that sound design programs focus on sound for moving image, but having a background in sound-for-image misses out on all the possibilities that can be created by sound design as "just" sound design.

An interview I conducted with a video game sound designer, Adele Cutting, made me think there may be a better way to teach sound design in schools, by focusing on sound before moving on to sound for picture. Cutting had been hired to design the sound for an audio-only video game: that is, a game designed with little to no visual component. She explained some of the differences when there's no image to design to:

> I worked on *Audio Defence: Zombie Arena* (Somethin' Else 2014), the audio-only game, the zombie shooter, and that's like the Holy Grail for a sound designer, isn't it? An audio-only game! It was a short turnaround—like four weeks—to do all the sounds. And it took me a good couple of days—probably three days—which is a lot of time when you've only got four weeks, to get my head around it. Because all these tricks that I did [with visuals]: Say, you were making a giant sound, you learn every time all these tricks to make it weighty and heavy. But when there's no giant's foot falling [visually], it didn't work. I really had to get my head around it. That there was no visual clue to hang on with, because I'm always talking about how audio fills in, how audio is the glue that holds everything together, and we fix things. We make things look better when the animator hasn't had time to do this, so we'll put a sound in, so nobody notices. We're always fixing things, and if things are far too slow, you can add audio and it speeds it up. You can add audio and make it go slower, but all of a sudden, [without visuals] it's just you. I found that game at the start very, very difficult because you have to be so focused. There can't be any fat on your sounds. It's just got to be the one thing that you need to hear, and you can't mix in [with visuals]. . . . I found myself chucking a lot of things out with the sound, to get the focus on it. . . . I felt it was so important that if there was only one sound going to be playing, or if you could only focus on one thing at once, it had to be the right thing. (quoted in Collins 2016, 119)

I am proposing that sound design, as a practice, may be better approached as an art form that stands alone from image, prior to learning about the complex things that happen when we put sound and image together. In other words, before we learn to put sound to image (looking and listening), sound designers are better served learning to just *listen*. I've designed this book based on my own teaching of sound design for about fifteen years now at several universities and in industry presentations and workshops, with the aim of helping others to structure a course in sound design beyond image. In an ideal world, students would then go on to learn sound for moving image in another course, and sound for interactive media in yet another course. We don't often get the luxury of teaching multiple courses on sound, however. Anyone studying visual production would get all kinds of courses in drawing, illustration, painting, printmaking, typography, digital arts, graphic design, and so on; sound designers rarely get that same kind of scaffolded and multifaceted approach to learning.

This book is about sound design as "just" sound design. I bring in examples from other media, but the many exercises I include are meant to focus the student of sound on just that—*sound*. But what does it mean to *design* sound? We hear the term "sound designer" applied to film or video games, but what exactly does a sound designer do? In fact, although the term is fitting, it was an almost accidental title. In the Hollywood movie system, a *sound editor* was (and still is) the person responsible for creating and selecting sounds for film (by substituting, eliminating, and adding to the original live recording or creating the sounds in postproduction). The term *sound designer* was first used to describe the work of Walter Murch. Director Francis Ford Coppola recalls:

> We wanted to credit Walter for his incredible contribution—not only for *The Rain People*, but for all the films he was doing. But because he wasn't in the union, the union forbade him getting the credit as sound editor—so Walter said, "Well, since they won't give me that, will they let me be called *sound designer*"? We said, We'll try it—you can be the sound designer. . . . I always thought it was ironic that "Sound Designer" became this Tiffany title, yet it was created for that reason. We did it to dodge the union constriction. (quoted in Ondaatje 2002, 53)[1]

Although the term *sound design* is most commonly associated with film and more recently video games, it is also applied to radio, theater, product design, and more. Traditionally, the goal of product sound design has been to reduce or remove sound, by engineering products that absorb (incorporating foam, perforations, etc.), block, or enclose the sound. Today, however, a growing awareness of the important role that sound can play in products is redefining the role of a product sound designer. Product sound design now has many of the same concerns as film and game sound design, that is, driving our emotions, rather than strictly information.

Increasingly, sound designers are finding a role in the growing audio-based media world of podcasts, smart speakers, and audiobooks. We can also add to sound design the growing field of sound art, in which artists use sound to convey their thoughts and feelings and express themselves, much as they have done for millennia using visuals. Artists are no longer confined to canvas; they can create multimedia works that incorporate sound or make sound the primary focus of their work.

---

1. In fact, Coppola misremembers, as Murch is credited as creating the "sound montage" in *The Rain People* (1969), and was first given the title of sound designer for his work on *Apocalypse Now* (1979).

Sound design can take place at the level of a single discrete sound or at the level of an entire soundscape. Tomlinson Holman, the inventor of the THX sound format, provides a succinct definition of sound design that will suit our purposes: "getting the right sound in the right place at the right time with the equipment available" (2002, 26). Of course, describing what is the *right* sound is a more complicated process that requires further exploration. Sound designers must work within the constraints of context, in addition to budgetary and technical constraints. But there are additional elements that must be satisfied: the aesthetic choices made will affect the overall reception of the work or product. Is it pleasing? Is it annoying? Designers must make choices about sounds based on the ways in which they want the audience to (consciously or unconsciously) interpret the sound.

Designers design sounds by:

(1) Choosing recorded sounds which, by their selection, context or combination, create something new. For instance, how sounds are juxtaposed—or the situational context in which they are used—influences their perception. This task may also include using unusual materials or getting unusual sounds out of everyday objects. Selecting or recording the right sound is an important design decision. Gregg Barbanell has talked about how he uses everyday objects for the sounds of gruesome bone-breaking in *The Walking Dead* TV series (2010–): "For 'breaking bones,' big, full stalks of celery are employed—not merely individual stalks, mind you, but *huge* bunches capable of producing layered, complex snaps. They give you this huge, sinewy stringy sound. . . . It's very effective" (quoted in Eddy 2015).

(2) Layering or combining several sounds to create a new sound, or splicing sounds together to create a new sound. For example, Ben Burtt describes the creation of the lightsaber sound for *Star Wars* (1976):

> I was a projectionist, and we had a projection booth with some very, very old simplex projectors in them. They had an interlock motor which connected them to the system when they just sat there and idled and made a wonderful humming sound. It would slowly change in pitch, and it would beat against another motor—there were two motors—and they would harmonize with each other. It was kind of that inspiration, the sound was the inspiration for the lightsaber and I went and recorded that sound, but it wasn't quite enough. It was just a humming sound, what was missing was a buzzy sort of sparkling sound, the scintillating which I was looking for, and I found it one day by accident. I was carrying a microphone across the room between recording something over here and I walked over here when the microphone passed a television set, which was on the floor, which was on at the time without the sound turned up, but the microphone passed right behind the picture tube and as it did, this particular produced an unusual hum. It picked up a transmission from the television set and a signal was induced into its sound

reproducing mechanism, and that was a great buzz, actually. So I took that buzz and recorded it and combined it with the projector motor sound and that fifty-fifty kind of combination of those two sounds became the basic lightsaber tone, which was then, once we had established this tone of the lightsaber of course you had to get the sense of the lightsaber moving because characters would carry it around. They would whip it through the air. They would thrust and slash at each other in fights. And to achieve this additional sense of movement I played the sound over a speaker in a room. Just the humming sound, the humming and the buzzing combined as an endless sound, and then took another microphone and waved it in the air next to that speaker so that it would come close to the speaker and go away and you could whip it by. And what happens when you do that by recording with a moving microphone is you get a Doppler shift. You get a pitch shift in the sound and therefore you can produce a very authentic facsimile of a moving sound. And therefore give the lightsaber a sense of movement and it worked well on the screen at that point. (Burtt 1993)

(3) Altering sounds through analog or digital signal processing, such as morphing sounds together (as in ring modulation); time domain effects (phasing, flanging); compression and limiting; reverberation and echo; and so on. For instance, the sound of the disc flying through the air in *Tron* (1982) was "a combination of a monkey scream backwards processed through a flanger and it was also another one of those weird synthesizer effects that I was able to create through the modulator, and also I took a big wire cable spin and that was the whooshing element. . . . I turned it [the monkey scream] backwards and you couldn't recognize that it was a monkey scream really" (Petrosky n.d.).

(4) Synthesizing a sound, or creating a sound based on granular aspects that are recombined from other sounds. For instance, consider the THX sonic logo, known as "Deep Note," created by Andy Moorer:

> I set up some synthesis programs for the ASP [synthesizer] that made it behave like a huge digital music synthesizer. I used the waveform from a digitized cello tone as the basis waveform for the oscillators. I recall that it had 12 harmonics. I could get about 30 oscillators running in real-time on the device. Then I wrote the "score" for the piece. The "score" consists of a C program of about 20,000 lines of code. The output of this program is not the sound itself, but is the sequence of parameters that drives the oscillators on the ASP. That 20,000 lines of code produce about 250,000 lines of statements of the form "set frequency of oscillator X to Y Hertz." . . . The sound was produced entirely in real-time on the ASP. (Whitwell 2005)

This book focuses on the first three of these four means to design sound. The programming and use of synthesizers to create sounds is a fascinating topic, but it requires at least a book of its own as well as more advanced skills. Likewise, interactive sound also requires a separate book to understand the complexities and software involved.

The aim of this book is to provide a set of material that, with each chapter, builds on previous work that you have learned and put into practice. I have interwoven theory and suggested further reading and listening materials throughout, with the hopes that you will take it upon yourself to improve your skills by exploring the many resources available to help you to learn about sound. I have suggested exercises to help you put the theory into practice, and while you may not want to complete all of these exercises, I believe that the more you undertake, the better you will become. Most of the exercises you can do on your own, so there is no need to be enrolled in a class to do these exercises, but a handful of exercises are better experienced with the participation of a partner or class.

In my experience, many introductory books on sound can get very technical with lots of equations and physics, which might put off a beginner coming at the field from an artistic background. It's my goal to focus on the creative side of sound design, and give you just enough of a technical foundation to get you started so you can put your creativity to work. Learning more about the technical side is an important step in a professional sound designer's training, but in my opinion that can happen after you begin to feel comfortable with the terminology and tools available.

I use Audacity as the software sound editor for the examples that demonstrate the techniques in this book. The reason for this choice is simple: it's free. Audacity has its limitations, and if you're serious about sound design you'll find yourself outgrowing it quickly, but if you're just dipping your toes into the waters of sound design, it's a great cross-platform tool or complement to other tools in your digital audio workstation (known as a *DAW*). It's important to note that Audacity is designed as a sound editor, rather than a *multitrack* editor. It's great for editing individual sounds, but as we'll see, the software becomes more problematic when dealing with mixing multiple tracks. The exercises can be undertaken in any other audio software you are comfortable with, like Audition, Logic, ProTools, or Reaper.

As well, there is a companion website to this book at studyingsound.org that provides examples, tutorials, some of the reading material, links, videos, and other resources that you can consult as you travel on your sound design journey. I'd love to hear about your successes with the exercises, and if you'd like to share your work, be sure to keep in touch!

The book is arranged in a scaffolded fashion: ideally you should follow sequentially, so that you can build on your skills as you go. In chapter 1 we start with learning to listen, and begin to think about sound in a new way to train those ears to do what they were born to do. In chapter 2 we'll begin to develop a language to talk about sound as an acoustic phenomenon, and learn the basics of digital audio. Then we'll turn our attention to recording and learn the basics of microphones in chapter 3. Of course, sounds don't exist in a vacuum, and the space in which sounds occur is important—and the focus of chapter 4. We'll begin to explore digital audio effects that mimic spatial effects, and then take a deeper dive into some of the other types of audio effects in chapter 5. Chapter 6 puts those technical skills together in exploring the theory and practice of mixing. In chapter 7 we'll explore an overview of spatial sound, or "3D audio." Chapters 8 and 9 take a different tack, and present some of the useful theories for understanding sound design and putting those into practice in creating sound for story.

Setting Up

To get started, we need to set up our listening space and our software. Try to find somewhere quiet to work in. While an increasing number of sound designers use home studios, it's important to minimize external noises in any audio workspace. If you have the option, choose a bigger room, which will generally sound better than a smaller room, but it's important that this room is as far from neighbors or external noise as possible—a basement is generally the quietest. Ideally, you can use some basic acoustic treatment to help reduce reverberation: foam and/or acoustic tiles can be placed strategically around the room. A blanket can be hung over a window to cut down some of the noise (and the window should be closed). A quick search online for "home studio on a budget" can provide some useful tips for your particular space and link to items you can purchase in your country to improve your sound experience.

If you are on a budget and don't have a private space to work in, a good pair of headphones is an important investment to listen to your work, and headphones are suitable for someone just starting out. Eventually you will want to purchase a good pair of studio monitors, but for this book headphones will suffice. You want to get a pair of over-ear or on-ear headphones, rather than earbuds. This is for both the sake of comfort as well as audio quality: the smaller the transducer, the less ability the headphones have to reproduce lower frequencies. Although wired headphones will give you better quality, wireless Bluetooth technology is improving. Wireless technology can be subject to electrical interference, however—I used to pick up the dispatch from the local fire station on mine! I keep mine plugged in when I work. If you can afford to purchase new headphones, do some online research as to the current best options for your budget.

As mentioned above, we are going to be using Audacity. To install Audacity, the software can be downloaded from the company's website, https://www.audacityteam.org.

In addition to the basic software, throughout the book you'll need to install some extra plugin modules. The instructions for downloading and installing plugins, and links to the plugins available, can be found on Audacity's website as well as studyingsound.org.

We will also need some basic recording equipment for the exercises. I recommend using an external microphone with your recorder (particularly if you are using your phone as your recording device). You can always purchase more expensive equipment later, but a basic kit is much cheaper than it used to be. A simple handheld recorder like the Zoom H4N or Tascam DR-40 is under $300. You want to get a recorder that has at least one XLR input and one line input, so that you can plug in an external microphone (professional microphones typically use XLR). These handheld recorders also usually come with built-in stereo microphones, but you'll find yourself using the external inputs more than the built-in mics. You can start with the built-in microphones, and, if you can afford it, add an external microphone—a shotgun microphone is a common first microphone to purchase for your sound design kit. As with much technology, the price of microphones has been dropping, and today the difference between a cheaper microphone and the high-end professional microphones can be minor in some cases. Certainly, while you are starting out and training your ears, the high-end microphones are not necessary. As your listening ability improves (and your playback technology also improves), you will notice the difference between the cheaper and the more professional microphones. Most sound designers have a selection of microphones for different purposes, but we'll come back to that later.

1 Hearing and Listening

What is the difference between hearing and listening? Lift your eyes from this page and look straight ahead for a moment: Notice that we see many more things than what we are looking at directly. We can focus on an object, but the eyes—and the brain—take in a lot more around us than just what we may or may not be aware of. I'm looking at my computer monitor, but I see my speaker monitors behind, and the posters on the wall behind them, and my shelves off to the side, and a second desk off to my left where another computer sits. My dog curls up in the corner. There is a black rug under my chair, and I see my arms moving, and all kinds of small details that aren't part of what I'm looking at. I *see* far more than I usually *observe*. A similar perceptual phenomenon happens with hearing. We are surrounded by sounds, and most of the time we are in a passive hearing mode, actively listening only when we are talking with someone (and actually paying attention to what they say!), in a potentially dangerous situation like crossing the road, or listening for a responding "beep" to a text message. What music we listen to is often just on in the background, a wallpaper of noise. Most of the time, sounds are just *there*, all around us. We listen with ears half open, not consciously paying attention to sound unless it's something that we are actively focusing on. We hear without listening, just as we see without looking.

Can we train ourselves to listen? How can we become better listeners? Like anything else we learn, what we need is practice, and we start our journey into sound design with becoming more aware of the sounds around us. We can learn, over time, to spend less time with ears half open and more time actively listening. Like a photographer walking around and mentally framing shots while scanning the landscape, we can learn to be aware of and thinking about sounds around us. Becoming a listener doesn't happen overnight, but with time and patience and practice, you will find yourself noticing more and more of the sounds around you. You'll find yourself hearing sounds that others haven't noticed, and you'll hear sounds that *you* never noticed before, and, sometimes, sounds you wish you hadn't noticed! Unfortunately, once you've opened your ears, the world becomes a very noisy place.

This chapter will introduce hearing and listening and begin to provide a language to think about and talk about sound. Listening is work that should be practiced and referred to again and again, until it becomes second nature. There are many exercises here to get you thinking about the sounds you're hearing, and training you to listen to them instead of just hear them. Training your ears is just like training your muscles in the gym: you can't transform yourself overnight. You have to keep going back and working at it, and it must be sustained or you'll find yourself losing your gains.

Exercise 1.1 Quiet Time

We've probably all tried sneaking into our house at night: every sound we made seemed suddenly amplified. Trying to *be quiet* is a great way to focus on actively listening. Try standing up from your seat without making any sound. Try it again with eyes closed. Listen to the sounds. How was the process of listening to your own sounds different from the way that you normally hear sound? (adapted from Schafer 1992)

1.1   Talking and Writing about Sound

Throughout this book you will find exercises and suggestions to get you to think about, experience, and practice sound in new ways. Keeping a notebook to write down your thoughts will help you to formulate your own ideas about sound and track your progress. You might also take a few moments to compare your own thoughts with those of friends, colleagues, or classmates as you follow along, or check the companion website (studyingsound.org) for another perspective.

Purchase a new journal for your sound practice. It helps if it's pocket-sized. You might wonder why I suggest a paper notebook and not your computer or phone. In theory, you could use a portable computer (laptop, phone, or tablet), but you'll find a pocket notebook will be handy to keep with you on a walk where you may not want to bring a computer (for instance, out in the rain). A phone isn't as effective to take notes on because the act of typing on a touchscreen requires you to focus visually on the phone and concentrate on that rather than the sound, which can interfere with the practice. You may also want to use your phone for other aspects of the exercises in the following chapters, as a pocket recorder, for instance, or to check frequencies or the volume of sounds you are hearing.

Once a day, practice sitting still for five minutes and writing down what you hear. You can sit in a different place, or sit in the same place. Sit at different times of day, and in different moods, or the same time and place and mood. What matters isn't so much what sounds you hear as your practice to actively attend to, concentrate on, and think about those sounds. You need to do this daily, rather than trying to pack in a week's worth all at once, because you need to start training yourself to listen, and this takes time. If you're serious about sound, listening is the most important skill you can have.

In addition to this daily exercise, keep writing down your thoughts about the other exercises, any readings or news media you come across that are related, as well as note any interesting sounds you hear in real life or in movies or other media, so you can reflect on your learning, and refer back to it in a few months and see your progress. You may also come up with some great sound design ideas that you don't want to forget, and your sound journal is a great place to jot these down as you go.

To be a sound designer, we need a language to talk about sound. Language is one of the tricky aspects of dealing with sound. And even after we've grasped the language, chances are we're going to have to talk about it with someone who hasn't yet learned that language! As children, we're taught a lot about visuals. We learn about shape and color and texture, and we learn the language to talk about these. If I asked you to draw a circle with a diameter of five centimeters and fill it in with a smooth, lime green color, you could probably come up with something very similar to what I have in my mind. But how do we talk about sound? Sound is time based, which makes it more difficult, and it's never the same twice. Even if we use an electronic reproduction of a sound, we don't hear it the same way twice, and the environment in which it's played is also always shifting and plays a role in our hearing. More importantly, we're also usually not taught a language to describe sound unless we are referring to musical sound, which has its own specialized language and doesn't actually refer to the *sound* of the notes played, only the notes themselves.

Exercise 1.2   Describing Sound

Undertake this exercise every day, and we'll build on it as we go: Take five minutes and sit quietly, writing down all of the sounds that you hear. The first time you try this exercise, you might come up with a list a little like this one, which are the sounds currently occurring as I type this out:

- Music in the background
- A car driving by

- My fingers typing on the keyboard

- Breathing of my dog next to me

- A scraping sound of someone shoveling snow outside

- The backup beepers on a truck at the construction site

- My own breathing

- Whirr of the heating duct

- Hum of the overhead light

This list is a good start, and we're training our ears based on how we've been taught to listen in the past; but let's dig a little deeper here.

1.1.1 Sounds and Their Causes

There are two ways I've described the sounds I heard in my listening exercise 1.2. The first is in terms of their *cause*—in other words, the thing that is causing, or making, the sound: for instance, "a car driving by." The problem with such a description when used to describe sound is that it only tells you what sound I'm hearing if you know what type of car is being driven (a truck sounds different from a Porsche), what the weather conditions are (tires in rain sound different from tires on dry road), what time of day it is (a car in the middle of the night will appear to sound louder), what the speed of the car is, what gear it is in, what the mechanical condition of the car is (is there a hole in the exhaust?), what kind of tires it has (winter tires make a different sound from summer tires), and more. Without all of this detail, we might conjure up a generic concept of "car-ness," but it's not a very accurate descriptor of what I heard. How the car is moving is an important indicator of what is happening: Are they squealing tires with some bass thumping out the windows, or are they creeping past very slowly, eerily, suggesting some form of surveillance or stalking—these are two very different sounds! We have to have an agreement on what my description of the car means to even begin to guess all of the associations with the sound of a car driving by.

Let's look at another from my list: "my fingers typing on the keyboard." We've all typed on a keyboard, but keyboards have very different sounds, and the speed of typing depends on the skill of the typist. The volume of the typing might depend on whether or not the person is frustrated or angry. The tempo may be altered if they are stopping and thinking about what they are typing, or if they know what they are going to type in advance. An Apple keyboard with its low-lying keys sounds very different from a cheap PC keyboard. I have one key that sticks and requires me to hit it harder. So again, "typing on a keyboard" is not really an accurate description of what "typing" sounds like, only the *cause* behind the sound. The first thing we can learn as sound designers is to be more descriptive in our journals. Moving forward, as you practice listening, *get as descriptive as possible* for each sound. This requires us to really concentrate on the many attributes that go into the sound, rather than just the cause behind the sound. Concentrate and think about the sounds you hear and imagine trying to describe them in a way that someone could use to reproduce the sound.

1.1.2   Onomatopoeia

The second type of description I used in exercise 1.2 relates to *onomatopoeia*: a word formed from the description of the sound it makes. I've used the "whirr" of the heating and "hum" of the light. We use these types of descriptions with animals a lot: a dog's "bow wow," for instance. But did you know that onomatopoeia is dependent on language and culture? A dog says "av-av" in Serbian and "hong-hong" in Thai. So much for using words to describe sounds! If you're a gamer or anime fan, you've probably heard the phrase "doki-doki": this is the Japanese term for the heart beating quickly. It's not just a literal sound, but also carries the meaning that one is in love, and their heart is racing. The Japanese actually separate onomatopoeia into three categories, and about 1,200 Japanese words are cases of onomatopoeia, compared to English, which only has about 400 (Kincaid 2016).

Exercise 1.3   Gerald McBoingBoing

Dr. Seuss created a character called *Gerald McBoing-Boing* (TV series, 1956) who talked in onomatopoeia sound effects: "*When Gerald started talking, you know what he said? He didn't speak words—he went* boing boing *instead!*" The animation uses sound effects, but the book relies on onomatopoeia to describe the sounds. How many onomatopoeia words can you describe off the top of your head? How much can you communicate with just onomatopoeia? Try to write an entire day's journal entry just using onomatopoeia (hint: you're going to have to make up some new examples of onomatopoeia).

1.1.3   The Importance of How We Think and Talk about Sound

How can we describe sounds in a way that everyone understands? To do this, we need to learn more about acoustics and use a more precise technical language for sound. There is so much more to sound than what was in my list in exercise 1.1. If you look at my list, you'll see I didn't describe where sounds were occurring in space in relation to my position: was the music in front or behind me? Did the car drive by me on the right or the left? How far away was the construction site? The placement of sound in relation to our own bodies also affects the way we perceive sound. We'll be tackling a language to describe sound and focusing our ears on all of these issues in the coming chapters. Gradually, as we progress through our journey, we will learn to fine-tune our descriptions of what we hear. For now, start to think about how descriptive you can get about the sounds you hear on your daily listening practice. Try to capture as much information about the sounds as possible. The more descriptive you get, the more you'll find you need to really focus on the sound itself and not just the cause of the sound.

How we think about and talk about sound influences how we use sound in our creative processes. In *sound libraries*—collections of sound effects recorded for our use as sound designers—sound effects are often categorized based on what caused the sound: "airplane" sounds, for instance, or "bird sounds." But sounds could be categorized based on what we might use them for: "scary sounds" (hawks or crow sounds are often used in horror), or "morning" sounds (the rooster or the dawn chorus). When we design sounds for media, we often use sounds that are not tied to their actual causality. For instance, we use the snap of frozen celery sticks for the breaking of a bone. Who would think to look for "vegetable sounds" in a sound effects library for their horror film unless they were aware of these uses?

In other words, how we describe sounds to ourselves and to others can influence the creative uses of those sounds. It's important, then, to think "outside the box" in our descriptions and categories, and to move beyond causality into other aspects of sound. To do that, we need to practice listening, and we need to learn a new language for talking about sound.

Exercise 1.4   Categorizing Sound

Take a list you created in one of your daily listening exercises, and think of the ways in which you might categorize these sounds. For instance, you might divide the list into opposing elements:

natural—human-made

pleasant—unpleasant

quiet—loud

rough—smooth

low—high

discrete—continuous

near—far

What other categories can you come up with to group your sounds? What do the categories tell you about the types of sound that you hear, and the ways that you think about sound?

Exercise 1.5   The Sound Walk

Sound walks are simply walking while paying attention to sound, rather than sitting in one place, so that we can experience several different places and listen to the changes. On a sound walk, we are quiet and listen with attention to all of the many sounds that we normally ignore. We can do sound walks alone or with a partner who guides us, blindfolded, around the walk. The sound theorist Hildegard Wester-kamp (2007) suggests first starting with listening to your own body while moving. Listen to your footsteps and how they change on different surfaces. Make a sound by clapping your hands or whistling. Try the sound in different rooms. How does it change? Once you've practiced listening to yourself, pay attention to the environment. Do you hear other people? Can you detect rhythms? What are the loudest and quietest sounds that you hear? Focus on a sound and walk toward it. Notice how it changes with proximity. Move indoors. How does sound change in different environments? How did your listening ability change when you couldn't see?

Exercise: 1.6   Destined to Repeat

"In Zen they say: if something is boring after two minutes, try it for four. If still boring, try it for eight, sixteen, thirty-two, and so on. Eventually one discovers that it's not boring at all but very interesting" (Cage 2013, 94). Find a sound that at first might seem boring, but after repeated listening becomes much more interesting. How does the sound (appear to) change over time? Describe it!

Exercise 1.7   Listening for the First Time

Listen attentively to something that you typically hear but never listen to, such as the full cycle of a dishwasher, washer, or dryer. What did you hear that you never noticed before? How difficult was it to pay attention for such a long length of time? Did you mentally add beats, or musical notes, or anything to force a structure or pattern onto it? How long were you able to listen before your mind started wandering? Can you train yourself to listen for longer? Repeat this exercise after you've finished the book, and compare notes with your first listen.

Exercise 1.8   Soundmarks

R. Murray Schafer, one of the first *acoustic ecologists*, writes, "Just as every community has landmarks which make it special and give it character, every community
will also have original soundmarks. A soundmark is a unique sound, possessing
qualities that make it special to a community" (1992, 123). Examples might be a
local public clock, foghorns, trains, and so on. Find and describe the soundmarks in
your community—either your home, your neighborhood, or the entire city.

Exercise 1.9   Sonic Fingerprint

What sounds are personal to you that others might be able to identify you by? For
instance, my dog used to be able to identify my car from all the others that went by
our busy street and would run to the window when he heard it. One exercise I try
in my classes is to have four students come up to the class with their sets of keys.
Facing the front, another student stands behind them and subtly shakes their keys.
Can the students recognize which set of keys is theirs by the sound alone? I find
the majority of the time they can guess their own keys, even though they've never
consciously paid attention to the sound before. Think about your own personal
sonic fingerprint(s): perhaps it's your car, an unusual walk, or your keys, and come
up with a list of sonic ways that someone close to you might be able to identify
you. (adapted from Schafer 1992)

Exercise 1.10   Sound Timer Reminder

Send yourself a little reminder to stop and listen. We can get distracted pretty easily
and forget to pay attention to what is around us. You can get a timer for your phone
or watch, and set it to go off a few times a day. When it does, take sixty seconds out
to focus on and listen to the sounds of wherever you are. Listen to how basic sounds
change depending on the environment—your footsteps change based on the temperature outside, what you're walking on, what mood you're in, what the weather is,
what other sounds are around you, where you are, and so on. Pick a sound to focus
on, like footsteps, clicking your fingers, or your breathing, and write down how that
sound changes throughout the day.

1.2   The Ear and the Brain: How We Hear

While we focus on listening practice, it's worth understanding what is happening on the biological side of hearing. In a sense, we hear with our whole bodies and not just with our ears. Our bodies have resonant cavities in them in which sound vibrates: our lungs, our bones, and even our eyeballs resonate with different frequencies. Scientists have tested the base human body resonance to be between 5 and 16 hertz (Hz) (Kitazaki and Griffin 1998). Different parts of our body vibrate at different rates, though, with our head vibrating between 20 and 40 Hz (Hz is a measure of vibrations per second: we'll come to that in chapter 2).

The human eyeball typically resonates at about 19 Hz, which is below the normal *threshold of hearing* (meaning we can't hear a sound at 19 Hz). In the 1980s, a scientist named Vic Tandy was working in a "haunted" medical lab that many people found left them feeling uneasy. One day he brought in his fencing sword and noticed it vibrating. He discovered that the sword vibrated at about 19 Hz, and traced the vibration to a fan in the building. Shutting the fan down shut down all the reports of ghosts. Tandy later tested the theory in a fourteenth-century "haunted" pub cellar and found the same frequency (see Jasen 2016). Could it be that what we call ghosts are just cases of our own eyeballs resonating? More recent work has found that the roar of a tiger is 18 Hz, and could be used to disorientate and paralyze prey in advance of an attack by resonating their eyeballs (American Institute of Physics 2000). Different frequencies of sounds, in other words, affect our physical body in different ways.

In addition to sensing sound through our bodies as a whole, a common means of hearing is through *bone conduction*, and hearing-impaired individuals can have some hearing sense through this method. It's been reported that the famous eighteenth-century composer Ludwig van Beethoven used bone conduction to hear after he went deaf, by using his jawbone. Clenching a wooden rod in his teeth and attaching it to the piano, he could sense the vibrations through his jaw (Larkin 1971). Bone conduction bypasses the eardrum, and vibrates the inner ear directly through the bones of the skull. Bone conduction headphones sit on the bone in front of (or behind) our ears, and are used by the military because they don't cover the ear canal, so can be used to supplement regular hearing for communication. In this way, we can hear everything going on around us with our ears, and any communication through the headphones. Apple was recently granted a patent for a method to incorporate bone conduction technology into their own headphones, so it's likely bone conduction is going to become more commonplace in the future (Dusan et al. 2013).

Exercise 1.11   Bone Conduction Headphones

If you're particularly interested in bone conduction, or want a set of headphones you can wear while also listening to the world around you (while jogging, for instance), you can purchase some bone conduction headphones for a reasonable price. If you have access to a pair, write down your experience of bone conduction listening to music or sound in your journal. How does the sound through bone conduction differ from regular headphone listening? What aspects of the sound are emphasized? Do you hear more or less through the bone?

Exercise 1.12   Bone Conduction with a Dowel

Here we will repeat Beethoven's technique. Get some wooden dowel (3 mm, or 1/4″ width is enough, at about 40–50 cm—15 inches—long) from your local hardware store. Put earplugs in your ears or use your fingers to block your ears. Put one end of the dowel in your teeth and bite down. Put the other end on a speaker, piano, guitar, or other vibrating surface. How does this alter what you hear? What aspects of sound do you miss out on?

Exercise 1.13   Bone Amplifier

In this experiment we will build a jaw-bone conducting amplifier (adapted from Oakland Toy Lab, n.d.).

Equipment List

Two wires, about 30 cm (10″) each

Wooden dowel, about 6 mm (1/2″) diameter, about 10–15 cm (6″) long

3.5 mm audio plug, or "mini jack" (male) for soldering—you may have to purchase these with a cap that you should remove

DC motor 1.5–3V 15K RPM

Wire stripper

Soldering iron and solder

Drill with 1/16″ bit

    Strip about 2 cm (1/2″) on each end of the wires.
    Solder one end of each wire to one of the tabs on the motor.

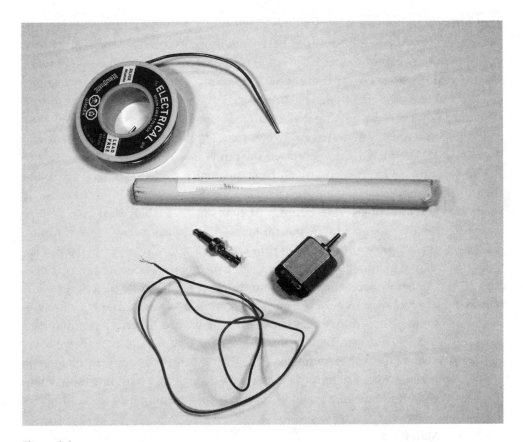

Figure 1.1
Equipment to build a bone conduction amplifier.

Solder the other two ends onto the tabs on the jack.

Drill a hole in one end of the dowel with the drill bit.

Push the end of the motor into the end of the dowel with the hole in it. You may have to wiggle it or put some pressure on it to get it to sit firmly in the hole.

Plug it into your computer, stereo or phone's headphone port and bite down on the dowel. You'll need to turn the volume right up, particularly if you're using it with your phone. Keep your fingers free from the dowel so it can vibrate correctly.

Put in some ear plugs or plug your ears with your fingers and listen!

1.2.1 The Outer Ear

While bone conduction is interesting, most of our hearing takes place through our ears. The outer, fleshy part of the ear is known as the *pinna* (plural pinnae). Another name for this visible part of the ear is the auricle. The pinna funnels sound toward our ear canal. If our ears were cut off, we could still hear, but it would be much more difficult, particularly in localizing (finding the direction of) sounds. With the pinna funneling sound into the canal, we can have a greater sense of our auditory environment and directionality. High frequencies reflect off the pinna in ways that differ according to the angle of the sound. Because we all have differently shaped ears, we hear sound slightly differently. In fact, our pinnae are so unique that earprint identification can be used in forensics like fingerprints (see, e.g., Meijerman, Thean, and Maat 2005).

Approximately 2 to 3 cm inside our ear holes—the *auditory canal*—is the eardrum, also called the *tympanic membrane* (tympani are kettle drums used in the orchestra). Unlike the skin of a drum, though, the tympanic membrane of our ear is a very delicate, thin membrane, approximately 0.1 mm thick. It can be easily pierced, which is why sticking anything into our ear canal—like cotton buds—is dangerous. The tympanic membrane is so sensitive that it can even be pierced by very loud sounds or pressure changes as when scuba diving or flying in an airplane. The many nerve fibers in the membrane make the eardrum very sensitive to pain. The tympanic membrane vibrates with the different sounds that enter the ear canal and transmits those vibrations through to the bones of the middle ear where they are amplified for hearing.

1.2.2 The Middle Ear

The middle ear consists of the space between the tympanic membrane and the oval window. This hollow space of the middle ear is known as the *tympanic cavity*, and is surrounded by the tympanic bone, which can function as a bone conductor. The tympanic cavity works as an amplifier that takes the vibrations from the tympanic membrane and transmits them to the inner ear via three tiny bones called the *ossicles*: the hammer (the malleus), the anvil (the incus), and the stirrup (the stapes). The malleus and incus bones developed through evolution from the upper and lower jaw bones in reptiles, which has been traced in fossil records. Our frequency range and sensitivity is determined by the shape and arrangement of these bones, which is why some mammals can hear ranges of sounds that humans cannot. The last of the three bones, the stapes, is situated inside a membrane-covered window in a bony separation between the middle and inner ear, known as the *oval window*.

The tympanic cavity is connected to the nasal cavity by the *eustachian tube*, which allows us to equalize pressure in our ears. We can manually adjust the pressure (for

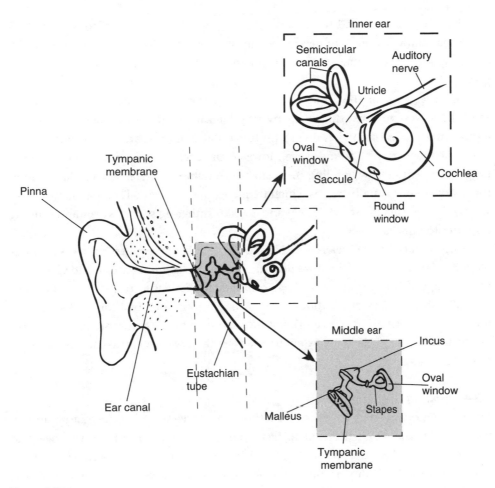

Figure 1.2
Diagram of the ear (not to scale).

instance if we are scuba diving) using what is known as the Valsalva maneuver, in which we pinch our nose closed and then blow out gently. Blowing too hard can damage the ear, so this must be done carefully.

1.2.3   The Inner Ear

When the stapes vibrates, it moves the fluids in the inner ear. Unlike other areas of the ear, the inner ear is filled with fluid and is responsible for both sound and balance. The inner ear contains the three components of the *semicircular canal*—three ring-like structures that are responsible for determining our sense of balance. As fluid, called endolymph, moves around the canals with the position of our head, sensors are triggered that allow our brain to determine our head position. Two vestibular sacs in the inner ear—the saccule and utricle—provide information about linear acceleration and gravity.

The other area of the inner ear, more important for sound, is the *cochlea*, a snail-shaped organ consisting of many tiny hair-like structures known as *cilia*. The entire length of the cochlea is lined with these cilia, and these are each attuned to different frequencies. As a sound wave moves in the cochlea, different frequencies will trigger different cilia by bending them slightly, sending electrical signals via the auditory nerve to our brain to tell us which frequencies were heard. There are many thousands of these cilia (between 12,000 and 24,000) gathering sound waves and sending impulses to our brain.

A large part of the cochlea is dedicated to the middle frequencies, with a peak range of 3500–4000 Hz. Most of what we hear in our world—including music and speech—is in this range of our hearing. In fact, when the gramophone was invented, most records (78 RPM shellac discs) until the mid-1940s had a top range of about 4,200 Hz (see Browne and Browne 2001). Even though we can hear much higher frequencies, most of our hearing takes place below about 8,000 Hz, and we are particularly sensitive to the speech range.

1.2.4 Hearing Development

The hearing organs start to grow in a fetus at just three weeks of pregnancy, and by week eighteen a baby will begin to hear sound. Soon after, the baby will begin to respond to voices or other noises it hears. Because there is a barrier between the baby and the world; the volume is muffled to about half of what we would hear outside the womb. But the baby can also hear sounds in the mother's body—the grumbling of the intestines, the heartbeat, and so on, and these would be heard much louder than by someone outside the body. Before a baby is even born, it can recognize the sound of its mother's voice. You may have heard of attempts to increase a baby's intelligence by playing music to it while in the womb, but there is no evidence that this works. What babies *do* learn, though, is the rhythm and cadence of what will become their native language—they can tell the difference between English and French, for instance, and can recognize the rhythm and pattern of stories that have been read to them in the womb after they are born.

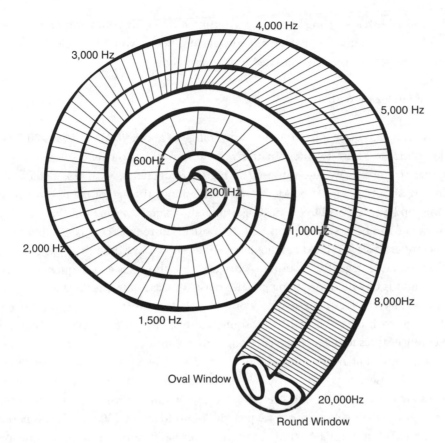

Figure 1.3
Diagram of the cochlea.

Exercise 1.14 Our First Sounds

Write in your journal what it must be like for a baby hearing sound from the womb. What sounds would they not be able to hear because of the muffled barrier of the womb? What sounds would they hear more loudly because of where they are?

Exercise 1.15 Hearing, not Listening

We hear constantly, even in our sleep, to the point where sounds can shape our dreams. Can you recall any dreams you've had where an external sound entered your dream? I know I've heard my phone ring in my sleep, then gotten up and discovered it hadn't rung after all. Try setting a timer on your computer or phone to play a sound quietly just before you wake up (before your alarm clock, if you set one), and see if it gets incorporated into your dreams.

1.3 Human Hearing Ability

Humans generally have a hearing frequency range of about 20 Hz to about 20,000 Hz (20 kilohertz, or KHz)—twenty vibrations per second to up to twenty thousand vibrations per second. As we age we lose the higher frequencies, with this deterioration beginning about age eighteen. Most people over about the age of thirty have already lost the top few thousand frequencies. Fortunately, there isn't much in that range that humans need to hear, so you likely will not notice. Currently there is nothing we can do to combat this age-related hearing loss. Some people have used this type of hearing loss to their advantage: the "mosquito" ringtone for phones is at a frequency of about 17 KHz, and is designed for young people to use (in classrooms, for instance) without the knowledge of older people, who won't be able to hear it. Older people have also used this to their advantage with "mosquito alarms," which are played outside some convenience stores to deter teenagers from hanging around.

Sound above our hearing threshold is called *ultrasound*. You're probably familiar with dog whistles: dogs can hear tones above our own hearing range, and most dog whistles are about 22 KHz. But even dog hearing is unimpressive compared to some other creatures: bats echolocate at frequencies of up to about 200 KHz. The wax moth can hear sounds as high as 300 KHz. On the other hand, some creatures can hear frequencies well below our hearing threshold, called *infrasound*—humpback whales have been recorded singing as low as 3 Hz, and the mantis shrimp, which can make sounds as high as 100 KHz, is also capable of sounds as low as 1 Hz.

Exercise 1.16 Imagining the Hearing of Others

We now know that some animals hear sounds we can't hear. But what's even more remarkable is the way that some animals hear. Some fish have cilia along a line on their sides, known as the "lateral line," so their whole body responds to sound waves. One type of squid, the longfin inshore squid, changes color based on sound—its chromatophores respond to changes in the environment, including sound. This exercise is a practice in creativity: imagine your sonic environment from the perspective of another creature, and write down what your listening environment sounds like.

Exercise 1.17   Online Hearing Test—Frequency Responses

You can test your hearing using a tone generator, which you can find at study-ingsound.org. Use headphones. Set the volume of your computer to a comfortable level. Start in the middle range, which as you learned when discussing the cochlea is not the technical middle, but the range where our hearing ability peaks, at about 3,500 Hz. Reduce the frequency to the point where you can no longer hear the sound. Record the lowest frequency that you can hear. Note that the low sounds may drop off because your headphones or computer can't reproduce those frequencies, not because your hearing is damaged. Now try going up in the other direction. What is the highest frequency that you can hear? A professional audiologist will test your frequency range for speech, but rarely tests above or below speech levels (in my experience, an audiologist tested only 200 to 8,000 Hz). You may need to use a subwoofer or studio speakers (monitors) to get a more accurate representation of your low frequency threshold.

Exercise 1.18   The Cocktail Party Effect

Most hearing tests will play multiple sounds at once to see how well you differentiate speech from other background sounds. The *cocktail party effect* is the brain's ability to focus on and differentiate sound in a noisy environment—like trying to listen to someone talking to you at a busy party. How loud can background sounds get before you can no longer hear what is being said? This speech differentiation is often the first thing many people notice if they have hearing loss.

When it comes to loudness, humans can hear sounds between 0 and 140 decibels (dB). Decibels measure perceptual hearing level, not loudness, so while there are sounds below 0 dB, we can't usually hear them with our ears (we'll come back to decibels in the next chapter). We can also hear sounds that are more than 140 dB, but it's painful, will cause permanent hearing damage, and will likely rupture our ear drum, so it's not practical for us to hear above that threshold.

We begin to cause damage to our hearing at about 80 dB if we're exposed to the sound for many hours, as in some workplaces, and the damage can build up over time. The European Union cutoff safety point for sound in workplaces is 80 dB. At 90 dB it takes less time for our hearing to become damaged, but for short periods of time 90 dB is usually safe. At 115 dB (which is quieter than many rock concerts!), even a very short sound will cause irreversible damage. The cilia in our ears *do not regenerate*, so once damage has been done, our hearing is permanently damaged.

Table 1.1 Approximation of sound loudness

180 dB	Rocket launch (measured on the platform)
160 dB	Gunshot (at close range)
150 dB	Fireworks
140 dB	Pain threshold
130 dB	Plane taking off
120 dB	Loud concert, yelling at maximum volume, siren
110 dB	Pneumatic drill, jackhammer
100 dB	Subway train, power mower
90 dB	Bass drum, legal limit for industrial noise in many places, motorcycle, loud club
80 dB	Busy restaurant, EU limit for noise exposure in workplaces without protective hearing
70 dB	Hairdryer, alarm clock, traffic
60 dB	Busy street, talking loudly
50 dB	Average conversation
40 dB	Mosquito near you
30 dB	Quiet room, recording studio background level
20 dB	Whisper
10 dB	Breathing quietly
0 dB	Leaf falling on the ground

Although we can't help age-related hearing loss, we *can* control *noise-induced hearing loss*.

1.3.1 Equal Loudness

Different frequencies have different perceptual volumes, since human hearing sensitivity varies with frequency. Lower frequencies drop off sooner, so low-frequency sounds are often given a boost by built-in equalizers in our stereo systems, to appear to balance out the frequencies. To demonstrate our hearing sensitivity, we can use what is called a *Fletcher–Munson curve*. These equal loudness contours measure decibel sound pressure level (dB SPL) over the entire frequency spectrum, providing a uniform appearance of loudness with pure sine wave tones.

To read these diagrams, first look at the bottom of the chart: these are the frequencies. Note that frequencies are not evenly spaced. This chart shows frequencies from 20 Hz to about 15 KHz. On the left are the decibels going up the chart. As stated above, we hear sounds of different frequencies at different perceptual volumes. So for a sound of 100 Hz, we would need a decibel level of nearly 40 dB to hear the sound. At 1,000 Hz,

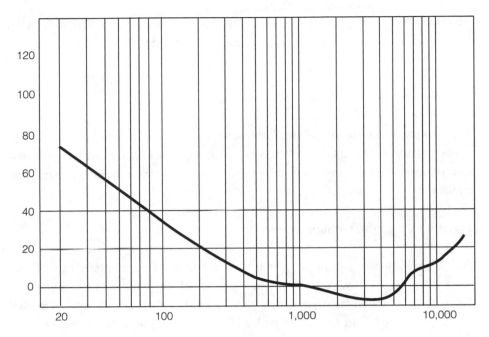

Figure 1.4
Fletcher–Munson curve.

where we are more sensitive, we can hear the sound at 0 dB, and in the range we are most sensitive to (about 3,000 Hz), we can actually hear below 0 dB in optimal conditions (it's unlikely you can hear these frequencies at that level anywhere but in a specially designed studio and only if you have excellent hearing).

In simple terms, our ears are not very good at hearing the lower frequencies compared to the higher frequencies. As loudness increases (the higher lines on the graph), the lower frequencies tend to flatten out as the level of volume increases. This means at higher sound levels the ear is more sensitive to (better at hearing) lower frequencies. Once we hit about 6,000 Hz the ear becomes less sensitive again. When we listen to music, we tend to "crank it up" because the added bass we can hear that comes with the volume increase means that the music feels richer, since we hear those bass frequencies more effectively at the higher volumes.

Exercise 1.19   Remanence

Remanence "is the continuation of a sound that is no longer heard" (Augoyard 2009, 87), like a musical earworm. The sound gives the impression of remaining after it's no longer there. Keep your notebook handy and track any remanence you hear in a day. Are there any common traits you hear among sounds that lead to remanence for you?

Exercise 1.20   Sudden Silence

Turn the power off where you live. How many sounds were there in the background that you hadn't noticed before? When you turn the power back on, how many new sounds are added back into your daily environment that your brain had learned to tune out?

Exercise 1.21   A Day without Sound

For this exercise you will need some equipment: at a bare minimum, a set of very good ear plugs. Ideally, you will use earplugs and then wear safety ear muffs over those. Remove sound from your life for one day (or half a day will suffice). Be sure you are going to be safe by staying with a friend or staying at home. Write a journal entry of your time without sound. Once you've spent a few hours without sound, how does returning to sound change the way you hear sound? What new sounds do you hear that you hadn't noticed before?

Exercise 1.22   Listening to Auditory Streams

In any soundscape, there are usually different things making sounds. We can think of these like different instruments in an orchestra. Find a busy soundscape, and spend a minute listening to each separate stream, or auditory source, focusing on the individual sounds and then on the whole. How many separate streams can you hear? What is the busiest place you've found in your sound walks? What is the least busy?

Exercise 1.23   Listening to Media

Once you have had some practice listening to a variety of natural environments, try comparing that with listening to a film or video game. If you're alone, it's easiest to do this exercise with a film, but if you have a friend with you who can play a game while your attention is on the sound, you can do it that way, too. Pick a film that you know well and have watched already at least once. Turn your back to the screen and just listen to the film. What do you hear that you didn't notice before? What sounds don't resemble the real world you've been listening to, and why?

1.4 Protecting Your Hearing

Probably the greatest damage to your ears is going to come from loud sounds, whether it's from your iPod, long-term exposure to a noisy workplace (everything from night-clubs and rock concerts to landscaping with power tools), or being exposed to a sudden loud sound. Fireworks at close range (150 dB), gunshots (140 dB), race car engines (140 dB), and industrial machines are the biggest culprits in urban life, but natural events can also cause great damage—thunder at close range is about 120 dB, and earthquakes have reached at least 250 dB. It's been estimated that Krakatoa was about 180 dB and ruptured the eardrums of people forty miles away. The Tunguska meteor explosion in Russia in 1908, the loudest known sound, was about 300 dB. Even the blue whale sings at nearly 200 dB! In other words, there are some things we can't control that we may be exposed to in our lifetimes, but most of the time we have control over the noise pollution by using earplugs (which usually reduce sounds by about 20 to 30 dB), ear protectors (a good pair will reduce sounds by about 30 to 40 dB), and not playing our music too loudly.

The canal from the outer ear to the tympanic membrane contains ear wax. The ear wax, called cerumen, may be wet and waxy or dry, depending on your genetics. The wax protects the ear from dust, microorganisms, and foreign material. Great caution should be taken when cleaning your ear wax with a cotton bud or other foreign body. It is better to wipe away any wax that has already exited the ear canal, and not put anything into your ears to remove the wax yourself. Not only do you run the risk of perforating your eardrum accidentally, but you can end up pushing the wax deeper inside, and impacting it inside your canal where it will reduce your hearing ability and must be taken out by a doctor with a special instrument. Don't use ear candles, tinctures, or medications to clean the wax unless you are under the care of a physician.

Cold weather can also cause damage to your hearing over time, so it's worth wearing a hat or earmuffs in the winter if you live in a northern climate. This damage is known as *surfer's ear*, a form of exostosis. It's not going to happen overnight, but over time the tympanic bone will thicken and develop new bony growths in an attempt to protect the inner ear from the cold. The thickened bone can actually trap water in your ear and lead to infections. If you spend a lot of time in cold water or outside in the cold, be sure to invest in something to keep your ears warm.

Tinnitus, often described as a ringing in the ear but which can also present as a hiss, a grinding sound, or other auditory phenomena, is often the first sign of hearing damage. Hearing damage can be caused by a number of factors: disease, injury, exposure to noise, stress, and medications can all affect hearing ability. Some common over-the-counter and prescription medications like acetaminophen, narcotics, antidepressants, and anticancer drugs can cause temporary or permanent damage to hearing, called *ototoxicity*. If you're serious about a career in sound, or want to protect your ears, it's important to discuss ototoxicity with your doctor and pharmacist whenever you are taking a new medication. Not all doctors are aware of the ototoxic effect of some medications, and you may not notice until it's too late, so do your own research. If you are taking medication and develop tinnitus, be sure to see a professional right away to discuss whether the cause could be your medication. In addition to diminishing your ability to work as a sound designer and to listen to music, studies have shown that hearing loss can contribute to dementia and depression, so it's worth caring about your ears!

1.5 Headphones Guide

Actively listening to sound is also referred to in sound design terms as *monitoring*, since you are monitoring what is occurring. Unless you have a home studio set up, headphones are the best way to monitor sound. A *frequency response chart* resembling the Fletcher–Munson curve graph is usually included on a box or leaflet with a set of headphones. Flat response is ideal for monitoring, but as long as you know where the peaks and valleys are on your headphones, you can make adjustments. For example, if you know your headphones have a bump at about 500 Hz, you can remember that when you go to listen to, mix or master files (more about that later).

There are several types of headphones to be aware of:

In-ear/earbuds: In-ear headphones, the earbuds that fit into your ear canal that come with your phone, iPod, and the like are cheap, portable, and convenient for jogging. They are often good for noise isolation—they filter out a lot of background noise. They are not good for monitoring, though, as there is little sound response in the lower frequencies. It's better not to use these except for convenience.

On-ear: On-ear phones are usually lighter and cheaper than over-ear (see below), and they are smaller, since they are designed to sit on, not over, the ear. These come in closed-back or open-back style. Closed-back (also called closed-ear or circumaural) headphones are sealed off to external sound, so you don't hear a lot from outside the headphones. Open-back headphones (also called open-ear or supra-aural headphones) are (as the name suggests) more open to outside noise, and there is more leakage to the outside world, but these are more comfortable for long periods and usually used in a studio for mixing and monitoring, where there isn't a concern about outside noise. These tend to be more expensive, as they have more expensive components and are the best quality overall. The sound quality is clear and there isn't a lot of leakage of sound, so a singer listening to a backing track while they sing might use this type of headphone. However, these can be heavy and less comfortable for long periods of time. Cheaper models tend to have a boost in the high-frequency range, which can become tiring to listen to. Bluetooth models allow for cord-free experiences, but interference can still occur, so it's recommended that you plug in while monitoring.

Over-ear: A decent pair of over-ear headphones, sometimes called *studio cans*, will serve you best for long periods of monitoring. Usually they will only have one cable attached to the headset, which is always on the left ear's side, to keep your right hand free of the cable for working (sorry, lefties). They also come in closed-back and open-back styles, and are good at isolating sound.

Noise-canceling: We don't use noise-canceling headphones for monitoring; these are convenient for travel when you have to block out steady noise like the hum of an airplane, but are not purposed for monitoring or music listening. We'll discuss how these work in the next chapter.

Reading and Listening Guide

Each chapter introduces some reading and listening suggestions. Take the time to read, listen, and answer the questions as well as add your thoughts about the readings or listening in your journal.

Michel Chion, the "Three Listening Modes" from *Audio-Vision* (1994)

Film sound theorist Michel Chion describes three ways of listening to sound. The most common is by identifying the cause of the sound, which we saw above in our own listening practice. He calls this *causal listening* (not to be confused with *casual* listening!). As we discussed, we usually talk of sounds in terms of the cause or source of the sound: a car motor, a bird, and so on. The second is *semantic listening*: this is what we do when we listen to people talking. The sounds are a part of a linguistic code, and we listen to the code as much or more than the sound itself. Chion draws on the *musique concrète* composer Pierre Schaeffer to describe focusing on the traits of the sound in *reduced listening.* To describe sounds in a reduced way we need a language to talk about sound, which we'll cover in the next chapter: we can focus on the texture, qualities, or timbre of a sound. We must also listen to a sound many times in order to separate out its acoustic properties from its cause. These three listening modes, however, fail to capture many of the other ways in which we may listen (and Chion acknowledges this). What other ways of listening can you imagine? Think about your listening journal and the ways you're practicing listening. Is listening to music different from listening to sound? Why or why not?

Pauline Oliveros, *Deep Listening: A Composer's Guide to Sound Practice* (2005)

Oliveros's book offers a different type of listening practice, focused on years of studying Zen and meditation. Oliveros reflects on her retreats and workshops and presents several deep listening exercises influenced by meditative practice. Perhaps most useful in my opinion is her slow walk, a meditation walk in which one attempts to walk as slowly as possible while listening. She tells us to "walk so silently that the bottoms of your feet become ears."

R. Murray Schafer, "I Have Never Seen a Sound" (2009)

Acoustic ecologist R. Murray Schafer explains his journey into studying the *soundscape* (analogous to landscape) of an environment. What sounds have been introduced to the soundscape during your lifetime? What would where you live sound like one hundred years ago? Five hundred years ago? What are the politics and power structures in the sounds of your environment?

There are many collections of soundscapes available, with a Spotify playlist linked on the studyingsound.org website. What do soundscapes tell you about the places you're listening to? What are the key sounds that differentiate them? Should we record our present soundscapes? Why or why not?

2 Sound

In the last chapter, we used some sound terms without fully defining them, including *hertz* and *decibels*. We commonly hear these words when talking about energy or sound, but many people use them without understanding them at all. In this chapter, we introduce the acoustics of sound. In most acoustics books you'll find lots of equations, but for the most part as a sound designer you probably won't be using any of them, so here we'll focus on the basics of sound without getting into the math.

2.1 What Is Sound?

Sound is caused by a vibration set in motion by some form of action that generates energy. Sound waves are often visualized as a ripple in water, as you would get from dropping a stone into the center of a pond or puddle, causing waves to radiate outward.

A slinky is a great tool to visualize sound waves. Hold both ends of the slinky, then send a "ripple" down the slinky—the wave travels down the slinky by moving one coil at a time, then each coil moves back to its original position. In the air, molecules move like the slinky: the vibration that causes a sound bumps the molecules one at a time in a wave. This is known as a *longitudinal* wave: the particles move in the direction parallel to the direction of the vibration of the wave. As the wave of molecules comes together, we call that *compression*, and as they pull apart we call that *rarefaction*. A wavelength represents one complete cycle of one compression and one rarefaction of the wave.

When we draw sound waves—or see them digitally represented—we don't use a longitudinal wave. Instead, we use a *transverse* wave. The transverse wave is represented as a *sinusoidal wave*, since the plotting of compressions and rarefactions of a longitudinal wave can correspond to the peaks and valleys of a sine wave. It's much easier to visualize sound as a transverse wave.

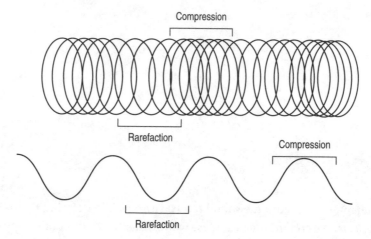

Figure 2.1
Compression and rarefaction in a wave (as illustrated by a slinky).

Exercise 2.1   Transverse Wave Model

This model is a transverse representation of a sound wave. You can build a sound wave model with some gummy bears, some wooden kebab skewers and duct tape. A 3 m wave requires about 60 skewers and 120 gummy bears.

First, set up two stands, one at each end of the wave you're going to make. It is also possible to get someone to hold them for you. The tape between the stands needs to be in tension. Run a strip of tape (e.g., duct tape) with the sticky side up between the two stands. Place skewers evenly, about four fingers-width apart, down the length of the tape. Optional: Run another strip of tape on top of the existing tape, sticky side down, to seal the skewers inside the strips of tape.

Stick gummy bears on the ends of skewers. You can use one per skewer-side to start: adding more will alter the speed of the wave. Stand at one end of the model, and lift a gummy bear at one end, then let it go. Notice how the energy comes back when the wave hits an end: this demonstrates how sound waves reflect. More pressure adds amplitude to the waveform.

If we stand and look at the wave as a cross section, we will see an approximation of the visual of a sound wave that we see in software. (adapted from Shaha n.d.)

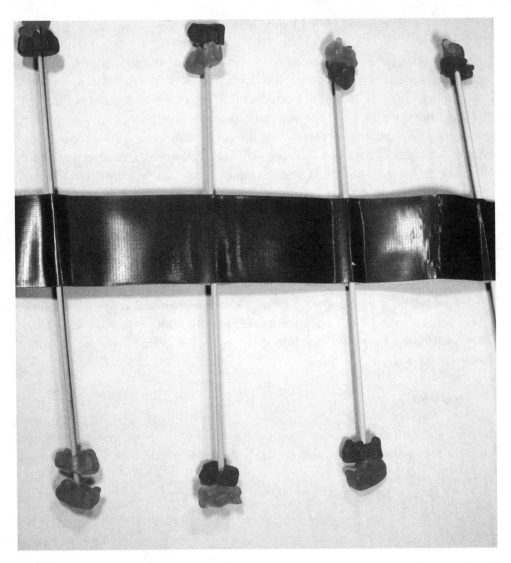

Figure 2.2
A gummy bear skewer sound wave.

Exercise 2.2   The Sound of One Hand Clapping

Zen *koans* are philosophical questions meant to make us think deeply. "What is the sound of one hand clapping?" is a classic one, as is "If a tree falls in the forest and there is nobody around to hear, does it make a sound?" These have no right answer, but are meant to have you think about the problem of answering them. Modern philosophers have also tackled the question of what sound is (e.g., Nudds and O'Callaghan 2009). Is it a property of an object, like a texture? Is it an object in itself, or is it an event? Since sound is just molecules moving until it touches our ear, which converts it to sound, can sound even *be* sound until we've heard it? Sit and ponder one of these koans and write your answer as a sound designer.

In Audacity, we can see the digital representation of a wave by generating and zooming on a sound. Select Generate on the toolbar, then select Tone. Click OK to select the default tone (sine wave, 440 Hz) in the pop-up window. You'll see a thick blue bar, which is the waveform (figure 2.3).

Now click on the zoom icon, or hit the F4 shortcut key, and keep clicking on the blue bar. You should end up after a few clicks with a nice sine wave (figure 2.4). The peak of the wave corresponds to the compression of the wave, and the valley corresponds to the rarefaction.

2.2   Frequency

The peak and valley of a wave combined is the wavelength, as we saw above, or one complete compression and rarefaction, and is known as a *cycle*. Each vibration of the

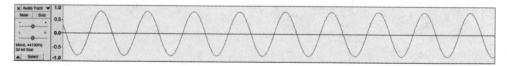

Figure 2.3
A sine wave viewed in Audacity.

Figure 2.4
A sine wave viewed in Audacity after zooming in.

sounding object causes one cycle, or one compression and rarefaction. The number of cycles per second is the *frequency* of the wave, commonly measured as *hertz* (Hz), which is named after Heinrich Hertz (1857–1894), but you may also come across *cycles per second*, or cps. You can remember *frequency* as how *frequently* the sound vibrates. Frequency is the pitch in musical terms: a higher frequency is a higher pitch. We use kilohertz (KHz) to measure in thousands of cycles per second: 1,000 Hz is 1 KHz. It's important not to confuse 1 KHz with 1 Hz!

We discussed in the previous chapter that humans have a range of about 20 Hz to 20 KHz, and we also found out that sounds exist at much higher and lower frequencies than that. Sound below our hearing threshold is *infrasound*. At the opposite end of the sonic spectrum, we use sounds that are higher than our hearing threshold for medical diagnostics (which is why the machines are called an *ultrasound*). These medical devices are in the range of 1 to 20 MHz (megahertz, or million hertz). But there is a reason why the radiologists need to press down on your skin to get an image: all sounds are absorbed by the air over time, and the higher the frequency, the greater the air absorption. A sound of 1 MHz would only be able to travel a few centimeters at best before it is absorbed by the air, so it's important for radiologists to get as close as possible to what they are trying to image to get the sound wave to your organs. We'll talk about this absorption and how sounds move in space in chapter 4.

As discussed previously, early records (78s) had an upper frequency range of less than 5,000 Hz. As you can see in table 2.1, most instruments make sounds only in this range. The only common musical sounds that you hear above about 7 KHz are the top end of the cymbals, snare and synthesized sounds, as well as harmonics (see below). For this reason, as discussed, what we call the "mids" or midrange in sound design is *not the middle of the human threshold*, but about 200 to 5,000 Hz. "Middle C" on a piano is just 261.6 Hz and the highest note is about 4,186 Hz.

Table 2.1 Approximate frequency range of common instruments

Piano	28 Hz to 4,186 Hz
Piccolo	523 Hz to 3,951 Hz
Trombone	82 Hz to 493 Hz
Kick drum	60 Hz to 150 Hz
Snare drum	100 Hz to 6 KHz
Soprano singer	262 Hz to 1050 Hz
Tenor singer	240 Hz to 525 Hz
Bass singer	87 Hz to 330 Hz

Exercise 2.3   Getting to Know Frequencies

You can practice learning approximate frequencies by using an app to help train your ears, or joining an ear training website, which will help you to focus on and learn to approximate frequencies. QuizTones, for instance, will test you on different tones so you can practice identifying them. Spend some time listening to different frequencies, and start trying to guess which frequencies you hear in your daily listening exercise.

Exercise 2.4   "The Prime Unity"

Spend a few moments relaxing in a quiet place. If you know how to meditate, you can try meditating for a few minutes. Hum a tone that appears to arise from the center of your being, what R. Murray Schafer calls the "prime unity." According to Schafer, in North America we often hum a B natural, but people in Europe tend to hum a G sharp. These are the notes roughly aligning to our electrical currents in our walls (60 Hz and 50 Hz), the sounds we hear all day long without listening. You may have hummed at another *octave*—music notes appear to repeat at other frequencies that are even ratios of the original, so you might have hummed at 120 or 240 Hz, and those would all be about a B natural. The Earth resonates at 7.83 Hz: was your hum a ratio of the Earth tone or the electricity around you?

2.2.1   Spectrograms

Spectrograms are useful for analyzing the frequency content of sounds. There are two ways to access a spectrogram in Audacity. Select your entire sine wave and select `Analyze: Plot Spectrum`. You'll get a plot that looks like figure 2.5.

Plot Spectrum is really useful when we need to know what frequencies are in our sound file. On the bottom are the frequencies—you'll see that they only go to about 10 KHz in this image. You can stretch out your window by clicking and dragging on the edge to view more frequencies. On the left are the levels. We'll come back to those soon. If you move your mouse to the peak of the file (the top of the mountain in figure 2.5), and look below the graph where it says "cursor," you'll see that the frequency peaks at 440 Hz, which is the frequency of the sine wave we generated.

More complex spectrograms are useful for editing audio files. We can switch to spectrogram view in Audacity by selecting the drop-down arrow beside the file name and change to spectrogram. Our sine wave at 440Hz looks like a solid gray band. It's possible to adjust the spectrogram settings in the spectrogram view in the same drop-down

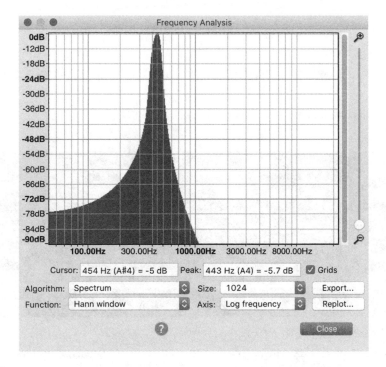

Figure 2.5
The Plot Spectrum window in Audacity.

menu where we just selected the spectrogram. I find Audacity's spectrograms to be harder to work with and prefer to use those in other software. There are also dedicated programs available without cost, such as Queen Mary University's Sonic Visualiser.

To give you a sense of how to read a spectrogram, it's helpful to look at a few different sounds with clear but changing tones, like sirens or whistling. Figure 2.8 shows a siren in Audacity and RX: the siren's sound rises and falls in frequency in an even pattern, with the loudest sound in the 700–1,500 Hz range.

It can take some time to get used to viewing and recognizing patterns in spectograms, but they are very useful for editing out unwanted artifacts in sound files. In the interview I recorded shown in figure 2.9, for instance, a police car went by. The siren is mixed in with the words of the speaker, so it's not nearly as clean as the siren in figure 2.8, but we can still see the frequencies in the file and edit them out to some extent. We'll come back to editing techniques later. For now, practice looking at some spectrograms by opening some sound files in Audacity and looking at them in the spectrogram view.

Figure 2.6
Spectrogram view showing the frequency at 440 Hz.

Figure 2.7
440 Hz wave in Izotope's RX. The brightness is an indication of the energy level (how high the volume is) in the file.

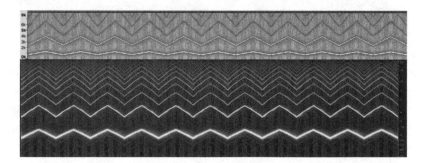

Figure 2.8
Spectrogram view in Audacity and RX showing a siren.

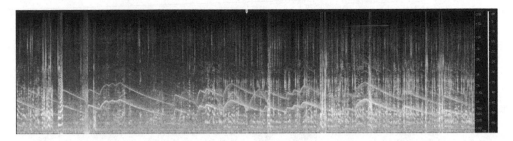

Figure 2.9
Spectrogram of dialogue with a siren in the background.

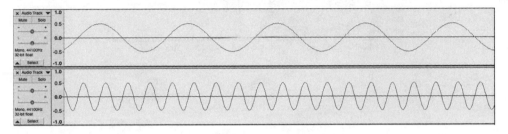

Figure 2.10
A 440 Hz sine wave (top) and 1760 Hz tone (bottom).

The spectrogram view in Audacity will show you the spectrogram of a single file, rather than the whole composition, as opposed to the Plot Spectrum window, which if you select more than one file will include them all in the spectrogram. To illustrate that point, let's add a second tone to our original file, this time at A6, which is 1760 Hz. Add a new file first. Select `Tracks > Add New > Mono Track` to insert a blank track below our first sine wave. Then select `Generate > Tone` and change the frequency to 1,760. If we zoom in on the two files, we can see that there are many more peaks in the higher frequency file (three times as many in the same distance) in the same amount of time, because there are more cycles in a higher frequency waveform (figure 2.10). If we plot the spectrum, we now see two peaks: one for each sound in the waveform editor (figure 2.11).

Let's mix the two sounds together. First, we need to reduce the amplitude, so select all, then select `Effects > Amplify`. Drag the slider to reduce the amplification by about 10 dB. When we reduce the amplitude of a sound, we call this *attenuation*. We need to do this because otherwise the new sound will be too loud (we'll come back to that later). Notice how the peaks of the waveform are now smaller. Now with both

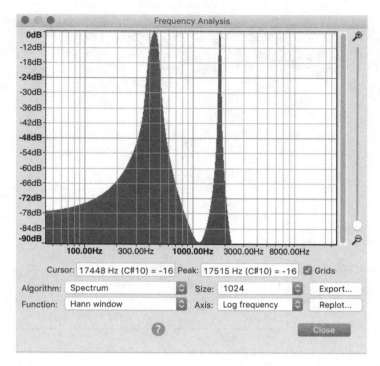

Figure 2.11
Plot Spectrum with two tones.

tracks still selected, choose `Tracks > Mix > Mix and Render`. This will mix the two tracks into one new waveform.

The new wave is a combination of both of the previous waves, but the sounds are still there. Select `Analyze: Plot Spectrum` again and we'll see the sounds there (figure 2.12). If we play the sound, it might sound like one sound: this is because the second frequency we chose is a tripling of the original frequency, in musical terms exactly two octaves above the original, which means they sound *in unison* with each other.

What happens if the tones are not in unison? Let's undo our project until we have two separate sound files and delete the second 1,760 Hz file. Now let's generate a new tone and enter 1,700 Hz. This frequency is not in unison with 440 Hz—it's not an even 3:1 ratio of 440 Hz. Listen to the two files together now, and notice how the tone is no longer smooth and clear, but has a harsher, more "buzzy" sound. These sounds are *dissonant*.

Figure 2.12
Two sounds now merged in Spectrogram view.

Exercise 2.5   Frequencies in Our World

Download a spectral analysis tool (or spectrogram) for your phone and head out for your daily listening exercise. Point your phone's microphone at some sounds and see what frequencies you are hearing. You'll notice that it's rare to get a single tone that makes it obvious what frequency you are hearing. This is because most sounds are made up of multiple frequencies. Looking at spectrograms can be a bit like looking at the code in *The Matrix* until you get used to it and learn to read them. Practice will make perfect! You will soon be able to guess the approximate range of fundamental frequency and their harmonics.

2.3   Consonance and Dissonance

As we saw above, some combinations of sounds feel pleasant to our ears, and others feel unpleasant. When two tones are within a critical bandwidth of proximity in frequency, they will sound rough, or *dissonant*. In music theory, some intervals (the distance between two frequencies) are considered dissonant, and others *consonant*. Consonant sounds are perceived as pleasant, and dissonant as harsh or rough. John Williams's theme from *Jaws* a great example of dissonance in action: the classic shark motif is a very dissonant interval back and forth (a minor second). Extremely dissonant were associated in medieval times with the devil, and were called "diabolus in musica," which is known as a tritone (the augmented fourth or flattened fifth interval). The interval has been used in rock and heavy metal, for instance, by Black Sabbath on the song "Black Sabbath" from 1970, or the first few notes of Jimi Hendrix's "Purple Haze." John Sloboda, a music scholar, explains (NPR 2012), "Our brains are wired to pick up the music that we expect, [and] generally music is consonant rather than dissonant, so we expect a nice chord. So when that chord is not quite what we expect, it gives you a little bit of an emotional frisson, because it's strange and unexpected."

Exercise 2.6   Consonance and Dissonance

Find a few examples of consonant and dissonant sounds in music you enjoy. They may be played one after the other (in *intervals*) or at the same time (in *chords*). Write down the song names and why you think the artists used those tones.

2.4   Amplitude

The height of the sound wave from the center of the waveform to the peak is the *amplitude*, which is the strength or power of the wave, and thus represents its volume: a higher wave is a louder sound. As we saw in the sound wave model experiment (exercise 2.1), increasing pressure (power) increases the amplitude. We measure the amplitude in *decibels*. The decibel gets its name from telephony—it's one tenth of a Bel (a deci-Bel, or *dB*, or sometimes *dB SPL* for "decibel sound pressure level"), named after Alexander Graham Bell. But loudness is a bit more complicated than that.

Sound can be measured most accurately in relation to the physical pressure level of the atmospheric change caused by a sound wave. This pressure is measured in pascals (Pa). However, we tend to measure sound according to weighted measures of loudness based on our perception. We learned in the previous chapter that perceptually we hear different frequencies at different amplitudes. For this reason, decibels are often recorded with what is called A-weighting, or dBA, which adjusts to those Fletcher–Munson curves. You should be aware that there are other weightings you may encounter, such as C-weighting (sometimes used in noise measurement), and Z-weighting (sometimes used in weather measurement).

Decibels work on a logarithmic scale, so every increase of 10 dB is an increase of 10 times the pressure level. If a sound is 10 dB, it's perceptually 10 times louder than 0 dB. If it's 20 dB, it's perceptually 100 times louder than 0 dB. If it's 30 dB, it's 1,000 times louder than 0 dB.

Exercise 2.7   How Loud Are Sounds? SPL Meter

Time to grab your notebook and head outside. First, download a sound pressure level (SPL) meter for your phone, such as Decibel Meter Pro or Decibel X. Go out and do your listening exercise, but this time have your SPL meter ready and write down how loud sounds are around you. Take a walk and pay attention to the loudness of sounds. Try to find a very quiet and a very loud place. What is the quietest place you can find? What is the loudest? Walk toward a sound and see how the decibels increase. The peak, sometimes referred to as Lpeak or Lpk, is the maximum value reached by the sound pressure without a time constant. It is the true peak of the pressure wave, and is usually C-weighted, not A-weighted. Note that smartphone apps are not as accurate as professional SPL meters, but they are fine for our purposes.

2.4.1 Digital Decibels

You may have noticed in your audio editor by now that digital recording goes *up to* 0 dB at the maximum and down from there into negative digits. 0 dB, in this case, is also known as 0 dB full scale (0 dBFS), and is the maximum that can be output before clipping or distortion occurs, so dB is a reference that is *relative*, not a real amplitude. When we work in digital sound, we use dBFS, or dB in negative numbers. This negative number scale is because the actual volume that we hear depends on how loud we turn up the hardware attached to our software; we can't always know what the hardware settings are set at, so we have to place sound relative to any hardware.

2.4.2 Clipping

It's possible to switch Audacity over to showing decibels instead of its –1.0 to 1.0 scale: to do so, select the drop-down menu beside the file name and select `Waveform (dB)`. You'll notice this distorts the waveform, so I'll leave it in relative scale for now. Create a sine wave as discussed earlier. Select it, and increase the amplitude by about 15 dB (`Effect > Amplify`). This time, we're going to tick the box that says "allow clipping." Don't listen to the sound without turning the volume down on your hardware! What happened to our nice smooth sound wave (figure 2.13)?

Zoom in on the waveform and you'll see that the top and bottom have been *clipped* off. When *clipping* occurs, the sound gets distorted and "crunchy"—it's gone beyond the peak, and this is why it's sometimes also called "peaking." It's only possible to recover some parts of a file from clipping with specialist tools that estimate where the sound wave should go, but avoiding clipping in the first place is always the best choice. Clipping—whether in recording or audio editing, is almost never a good thing, which

Figure 2.13
Amplitude increased to clipping.

is why the Amplify window has that tick-box for "allow clipping." If we undo our amplification, and go back to the original wave and untick that box and now attempt to boost the signal too high, the OK button will be grayed out until we reach a volume that doesn't clip.

Exercise 2.8 Amplitude in Composition

Find one sound and place it at different amplitudes in a *composition*, or collection of sounds. What do you learn about hearing the sound at different volumes? Download some loud sounds and put them into a sound composition and make them quiet. Then do the inverse: find some very quiet sounds and make them very loud. How does the change in volume alter your perception of them?

Exercise 2.9 The Sound of Silence

Not just a Simon and Garfunkle song: think for a moment about silence. Does it ever exist? How might we learn to use silence (or at least quiet) to give space to a piece, to add anticipation? Create a sound composition using only silence and very quiet sounds like pin drops or leaves falling. When we strain to listen, how does that change our relationship to the sound?

2.5 Timbre

Why do two sounds at the same frequency sound different? *Timbre* is the distinguishable difference between two tones that are the same frequency and loudness. Timbre is partially determined by the *harmonics* and *partials* of a tone. When objects vibrate, they tend to vibrate at a particular frequency—this is known as the *natural frequency* of the object. All objects have at least one natural frequency at which they vibrate. Some have a fairly simple sound wave: a flute or a tuning fork, for instance, tend to have close to a pure tone, or single frequency at which they vibrate. Most objects, however, vibrate at more than one frequency at a time. The dominant (loudest) frequency that you hear is known as the *fundamental*. Any sine waves that form part of the sound are called partial tones or *partials*. These can be inharmonic or harmonic.

When a string vibrates, the main sound you hear is from the vibration of the whole string back and forth. That is the fundamental frequency. But the string also vibrates in halves, in thirds, fourths, and so on. Each of these fractions produces a sound, called a *harmonic*, which is usually lower in amplitude than the fundamental. The string vibrating in halves produces the second harmonic; vibrating in thirds produces the third harmonic; and so on. If the fundamental is 60 Hz, then the harmonic order is 120 Hz, 180 Hz, 240 Hz, and so on. If you play a string instrument, you can sound a harmonic by placing (not pressing) a finger onto a string in a particular place and plucking the string: one of the harmonics of that string will play.

Complex sounds consist of many harmonics. The waveform for these sounds also visually looks more complex. We call the complex sounds that make up a tone harmonics if they are part of the harmonic series, which means that they are positive integer (whole number) multiples of the fundamental. They may also be called partials, but usually we reserve the term partials for when there are inharmonic sound components to a sound.

Let's take a look at a different type of wave to demonstrate. Under your sine wave, generate a new tone (Tracks > Add New > Mono, then Generate > Tone); this time select Square Wave under the waveform drop-down menu. We'll set it at 440 Hz, just like our sine wave. First, play the sine wave by selecting the solo button on the side of the track. Then, unclick that button and solo the square wave. Quite a difference, but they are both 440 Hz!

Zoom in on the waveform (figure 2.14). The square wave looks like its name—it is a square wave. Delete the sine wave, then select the square wave and plot the spectrum (figure 2.15). It doesn't look anything like the sine wave, because it has lots of harmonics. It's no longer just one tone like the sine wave; instead, it consists of lots of tones, at decreasing amplitudes. The square wave has harmonics at *odd* multiples of the original fundamental. If we look at our Plot Spectrum, we see the fundamental (the biggest peak) at 440 Hz, and the second biggest at 1,300 Hz, then one at 2,200 Hz, and so on. All of these other tones are the harmonics of the square wave.

All acoustic musical instruments have some harmonics. The quality of the harmonics is part of what gives instruments their unique sound. Some instruments have mainly odd or mainly even harmonics, and some have inharmonic partials. When it comes to electronically generated sounds like sine waves, square waves, or more complex synthesizer sounds, we can know what sounds are being generated by the

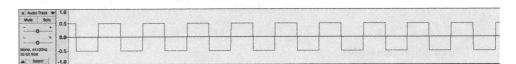

Figure 2.14
Square wave form in Audacity.

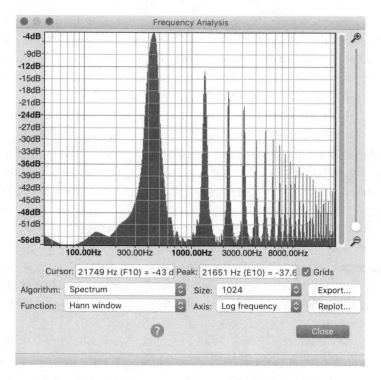

Figure 2.15
Plot Spectrum on a square wave.

algorithms involved in producing their harmonics. We already know a sine wave has no harmonics. A square wave has only odd harmonics. The other option to generate a wave in Audacity is a sawtooth. What are the sawtooth harmonics?

Exercise 2.10   Analyze Some Harmonics

Play various sounds into a spectral analyzer to determine the harmonics. Try, for instance, a tuning fork, a guitar, a xylophone, a flute, or a kalimba. What are the harmonics that you can hear by listening—and what does the spectral analyzer tell you?

2.5.1   Phantom Frequencies

When we hear the harmonics of a note without its fundamental, our brain might fill in the fundamental. This filling-in by the brain is known as a *phantom fundamental* or missing fundamental. If we know the ratio of the harmonics in a sound wave, we can give the listener the harmonics without the fundamental. So if we have a sound with a fundamental frequency of 100 Hz, we can produce sounds at 200 Hz, 300 Hz, 400 Hz, 500 Hz. If we have small earbuds, they can't produce sounds at 100 Hz, but if we play the harmonics, our brain can fill in the missing fundamental. The software company Waves has a plugin called MaxxBass that creates this effect, but we can also try and create it ourselves. The difference between real acoustic bass and the ghost fundamental is that we don't hear the vibration of the wave in our body the way we do with real low frequencies.

We can remove the 440 Hz sound from our sawtooth wave by using a notch filter (more about filters later). Generate a sawtooth waveform, use Effects > Notch Filter and type in 440 Hz. Now take a look at the Plot Spectrum (figure 2.16): 440 Hz is gone, but do you still hear it?

Another place we encounter frequencies that aren't really there, but are filled in by our brain, is with *combination tones* (in music, they are sometimes called *Tartini tones*,

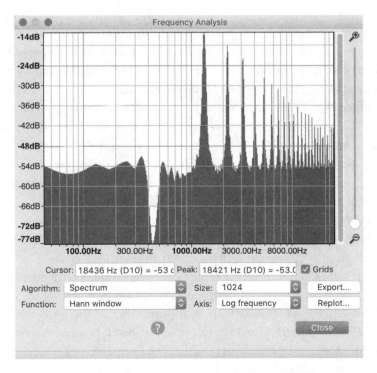

Figure 2.16
A phantom fundamental: the frequency will be mentally filled in by the brain.

after the violinist Giuseppe Tartini who frequently used them). When we hear pure tones like sine waves, our ears will sometimes hear a combination tone that we can perceive, faintly, in addition to the two frequencies that are actually there. *Difference tones* occur when two sine waves are played together, and the difference in frequency between those two frequencies is also heard. *Sum tones* are the two frequencies plus their sum. Create three new mono tracks in Audacity (`Tracks > Add New > Mono Track`). In the first, generate a sawtooth wave tone of 300 Hz. In the second, generate a tone at 1,000 Hz. In the third, generate a tone at 1,300 Hz. Solo the 300 Hz tone. This is the difference tone we're going to try to hear. Now mute the 300 Hz tone, and play the other two tones at once at a fairly high amplitude. Listen for that 300 Hz—the difference between 1,300 Hz and 1,000 Hz. Tartini used these extra phantom notes with his violin playing. The ear actually plays back these frequencies, in what has recently been termed *otoacoustic emissions*. That is, we don't just *hear* the sounds; our ears actually emit the sounds (see Connolly 2015)!

2.6   Wave Interference

Waves rarely exist in isolation. They're often meeting other waves in the air, and these will interfere with each other. Wave interference is important to understand because it is the basis of many processing effects we'll be looking at in the coming chapters. If two waves meet in air, if the waves are of equal wavelength they will combine, creating a kind of "superwave." This combination of waves is known as *constructive interference*. The result is an increase in amplitude. Create two sine waves at 440 Hz, both with an amplitude on the scale of 0.5 (this is not decibels, but the 0.1–1.0 scale used by Audacity). If we mix (`Tracks > Mix > Mix and Render`) the two waves, we get one new wave at the 1.0 amplitude (figure 2.17). We have created constructive interference across the entire waveform, since the two waves were exactly the same.

You probably learned in school that lights of different primary colors can mix to make secondary colors; for example, red and yellow make orange. When we mix the three primary colors (red, yellow, and blue), what do we get? White. Our minds may say it shouldn't work, but it does! The same kind of phenomenon happens with sound waves, and is known as *destructive interference*. When we have two of the same waveforms that are exactly *out of phase* with each other (that is, the peak of one lines up with the valley of the other), we get silence. So let's take our two sine waves in Audacity again, and this time shift the second wave using the time shift tool (or use the keyboard shortcut F5) to adjust the second wave to be out of phase with the first (figure 2.18). Then, mix and render the two files; we now get no sound (you may have to fiddle

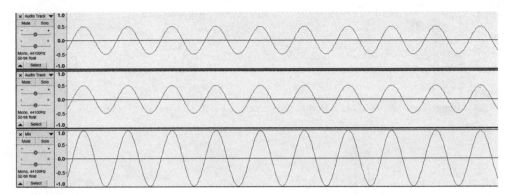

Figure 2.17
Constructive interference: two waves at the sample amplitude mixed result in a doubling of amplitude.

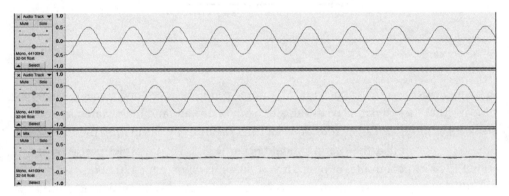

Figure 2.18
Destructive interference: line up two waves exactly out of phase and `Mix and Render`.

a bit to get the second wave exactly right)! This is how noise-canceling headphones work—they sample the incoming sound and create a destructive pattern to cancel those frequencies out.

Another wave interference phenomenon is beating. When two sound waves of very close frequencies are played at the same time, the result is what are called *binaural beats*. For instance, two sine waves one hertz apart will alternate between being in phase and out of phase with each other, producing a beating sound. The beating occurs when one frequency is compressing the air while the other is in rarefaction (expanding), causing this pulsing in the air. For instance, if you play one sound at 200 Hz, and another sound at 202 Hz, a noticeable beating will occur (figure 2.19). These usually only work

Figure 2.19
Binaural beats: two waves of close frequencies produce beating effect when mixed.

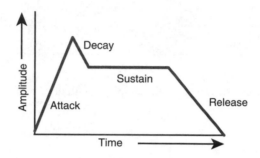

Figure 2.20
Sound envelope, or ADSR.

in frequencies lower than about 1,000 Hz. Some people believe—and some research has shown—that these binaural beats can alter your brainwave and brain states, increasing alertness (Lane et al. 1998). You can even time the beats by calculating the number of times per second that the beating should occur (e.g., 2 Hz = 2 cycles per second; 2 × 60 is 120 beats per minute, or bpm). If you set a metronome for 120 bpm, you should hear the beating in time with the metronome. This method is one way that piano tuners tune pianos, by calculating the frequency differences between two notes using the tempo of the beats.

2.7   Sound Envelopes

Another aspect of a sound's uniqueness that determines its timbre is the sound's *envelope*. The envelope represents how the *amplitude of the sound changes over time*. The envelope is sometimes referred to as *ADSR*, for attack, decay, sustain, release (figure 2.20).

The *attack* is how quickly the sound reaches full volume after the sound is initiated. The *decay* is how quickly the sound drops to the sustain level after the initial peak. The *sustain* is the constant amplitude that the sound takes after decay until the sound finishes. The *release* is how quickly the sound fades when a sound ends. Most envelopes are represented with these four stages, although some synthesizers give you five or six

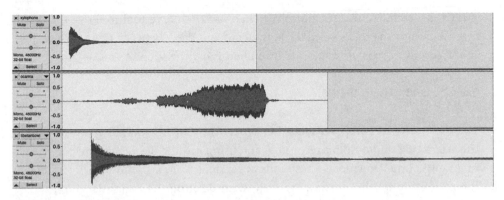

Figure 2.21
Comparison of some sound envelopes: a xylophone, an ocarina, and a Tibetan bowl.

stages. The attack portion of a sound is usually the most important, and often contains starting *transients*: tiny elements of a sound that are not harmonics of the note but are a result of the action that set a sound in motion.

Plug in a microphone and be sure to set the mic to be the input device. Hit the red record button and then clap near the mic. If we zoom in on the clap, we can see that it has a very fast attack, and a short decay, sustain, and release. Now let's try and record a sound that will sustain longer. Note that the envelope can vary a lot between different types of sounds: the attack isn't necessarily short and sharp, and the sustain or decay isn't necessarily long (see figure 2.21).

We can use the envelope tool in Audacity (underneath the top bar) to adjust a sound wave's envelope. Select the envelope tool and click on the sine wave. We can now adjust the envelope of the wave by dropping points onto the wave by clicking where we want to increase or decrease the amplitude.

Exercise 2.11   Reverse Envelopes

Reverse a few sounds (`Effects > Reverse`) and note how the envelope of the sound changes. How would you describe the sound you hear now?

Exercise 2.12   Attack Splice

Splice the attack of one envelope onto the DSR of another. To do this, select the point in the file where the attack peaks, then use `Edit > Clip Boundaries >`

`Split` (shortcut command/control I). Then use the move tool to move the decay away. Do the same to a second track, then cut and paste between them. Which part of the sound wave is most important to identifying the sound?

Exercise 2.13   Using the DSR

The experimental composer Pierre Schaeffer wrote in his journal about removing the attack of a sound from a recording, and using just the decay, sustain, and release as a new sound. Download ten sounds at random from a sound library and delete the attacks by splicing them using split, then deleting the attack, leaving just the DSR portion of the envelope. What do you hear in these parts of the sound? Can you create a new sound from just these portions?

Exercise 2.14   Exaggerated Envelopes

Find or make sounds with different parts of the envelope exaggerated (louder, longer, or more emphasized through timbre than through other parts). How does each aspect of the envelope change the sound you hear?

Exercise 2.15   Drawing Envelopes

Now that you understand envelopes, practice drawing what you imagine are the envelopes of sounds in your daily listening practice. If you can, record them (e.g., on your phone) and then check your accuracy until you feel comfortable.

2.8   Smearing, Rhythm, and Masking

Sounds are rarely heard in isolation, of course. The interaction between sounds also affects their perception. *Temporal smearing*—a blurring of distinction between or across sounds—can occur if sounds are too close together in time. Our brain tends to group sounds into higher-level patterns, or *rhythms*. Some researchers have suggested that these patterns are influenced by our experiences—particularly our native language (Iversen, Patel, and Ohgushi 2008). Have you found in your listening exercises that there are patterns that your brain imposes onto some sounds you know are steady and without pattern? Pay attention to these sounds in your next listening exercise and try to distinguish the actual rhythm from what your brain has imposed as rhythm.

Sounds can appear to be covered up by other sounds. When this happens we refer to the sounds as having been *masked*. If you have noisy neighbors you may find comfort in using a white-noise generator, which creates all of the frequencies in our hearing range at one time, blocking out some of the other sounds in the environment. Rain, waterfalls, and other similar sounds create a white noise effect that is useful for helping to mask sounds around you. In Japan, public toilets use a "sound princess" (*otohime*) button that plays a fake sound of a flush, to mask the sound of bodily functions in the public restroom. Many free sound systems available online are designed to mask noisy office mates, roommates, or neighbors.

Sharp, short sounds tend to stand out, even from white noise, in part because our brains pay more attention to them. Different frequencies will also cut through the noise in different ways. Those backup beepers on a truck are really annoying because they are designed to cut through all of the other sounds on a construction site with a particular amplitude, envelope, and frequency.

We looked at masking when we talked about the cocktail party effect in chapter 1 (exercise 1.17)—at what point did the noise in the background become too loud for you to focus on the vocals? In engineering terminology, we speak of this in terms of the *signal-to-noise* ratio, where the *signal* is the thing you want to hear, and the *noise* is the background noise that you don't want to hear.

Exercise 2.16   Signal to Noise

Open a short sound in Audacity. Then create a new track and use `Generate > Noise`. We can use the envelope tool to quickly adjust the level of the white noise in the background, and see at what point the signal becomes lost in the noise.

2.9   Selecting Sounds: Sound Libraries

As discussed in chapter 1, many sound libraries are available online or recorded on vinyl, CD or hard drive. These are often searchable using a variety of keywords. An online source for sounds you can use freely is freesound.org. Sound designers like to make their own sound libraries, but there will always be some sounds missing, and so other libraries can be very useful. Collecting, categorizing, and slating sounds (see chapter 3) is also very time consuming, so libraries can save us lots of time. Most professionally recorded libraries will have to be purchased, but for our exercises we can use free libraries.

Many sound designers hear commonly used *samples* (recorded sound effects) from libraries repeated often in media, and it becomes irritating. A particular fox screech from a BBC sound effects library, which I hear frequently on British television, immediately distracts me from the story because I've heard the exact sample used too often. Some audiences beyond us sound designers do pick up on this repetition after a while. An article in the British newspaper the *Guardian* commented on the repetition of the same horse sound in many programs (Hunt 2019). There is also a famous scream sound effect, the Wilhelm scream, that has been used in about 400 films and television shows; it started as an inside joke among sound designers but is becoming so commonplace now that many people are aware of it. A good rule of thumb is, be careful of where we are selecting our sounds, and if it makes sense, then alter those sounds in some way so as to not just "drag and drop" a sound effect from a library into our project.

Stephan Schütze, who records and sells sounds as the Sound Librarian and works as a sound designer, explained to me the process of recording and then using those sounds to design. I asked him to explain how sound designers use his samples:

> That I think depends a lot on the skill and the creativity of the sound designer. I can recognize a lot of my sounds, and the less they manipulate them, the more recognizable they are, but one of the other things that I do is that I really love working with sounds. I think that a good analogy would be a child in kindergarten doing finger painting. They get to get their hands into the paint and make a mess everywhere and then they can play around and they can get different colors and mix them all together.
>
> So one of the first things I do with any sound, if I get a raw sound and it doesn't matter what it is, I'll put it into an editor, and I'll usually pitch shift it up two, three, four octaves and have a listen to it, then I'll pitch shift it down, two, three, four octaves. Because what that gives to me is this is the scope of what that sound can be. And it teaches me things as well, because any sound being a raw tool, it's like getting a piece of wood. You and I will look at a piece of wood and think, "OK, it's a square block and we can do something with it." A real craftsman with wood will look at that and go, "Where does the grain go? What is the shape? What is the dimension of this? How can I shape this? What can I do with it?"
>
> . . . And so sound design obviously involved potentially pitch shifting the sound, but also blending them with other sounds [and] combining them, so the more you deform a sound, the more you combine it with other things, the less recognizable it will be from its original source, but I really think that understanding the full potential of what a single sound can do means that that whole craft of sound design is something that you can, you get so much more potential out of. (quoted in Collins 2016, 196)

We'll talk more about ways to alter sound in the coming chapters as we learn some processing effects. Schütze here mentions just one—pitch shifting—but there are many things we can do to manipulate a sample.

As we found in chapter 1, in media we sometimes categorize sound effects according to their use. Different libraries categorize sounds in different ways, which can be frustrating. There is no standard way to categorize sounds, and over time you'll come to develop your own ways to store categorize and organize your files. Most sound libraries will have some searchable tags, however, that relate to the causality of the sound.

When you were thinking about sounds in your daily listening, did you notice that some sounds had short envelopes and stood out, while others were long and didn't have a discernible beginning or end? You may have categorized these as continuous sounds. The mains hum or fan of an HVAC system in a building, for instance, may be a continuous sound: we can call these sounds parts of the *ambience* (not to be confused with ambiance!), sometimes called *atmos*, short for atmosphere, or beds, since they are laid down to support other sounds. Often we have layers built into the ambience—outside, for instance, we might have a babbling brook and some wind in the trees. Along with those ambient sounds will be *spot* sound effects, sometimes called *hard effects* or *one-shots*. These are single-source objects that are *discrete* sounds. Other sounds include Foley, named after sound effects artist Jack Foley and usually reserved for film and, more recently, video game sound. Foley refers to sounds, including door slams, footsteps, clothing rustling and so on, that are recorded by actors (Foley artists) performing them live to time the synchronization to the image. We also have *designed sounds*—sounds that have been manipulated with various digital effects—we'll come back to these later too. Designed sounds can also be synthesized sounds, but we won't be dealing with those in this book. Sometimes sound effects libraries are categorized into these uses of the sounds.

2.9.1 Copyright and Sound Samples

When we download a sample, we need to check the creator's copyright. Just as a photographer owns the rights to their own images, a sound recordist owns the copyright to their sounds, and can choose how we can use their sounds. Most sounds on freesound.org are licensed under various creative commons licenses. They may be in the *public domain*, which means we can use them however we like. They may be *attribution no derivatives*, which means we can use it in a project, but not if we alter the sound in any way, and we need to give them credit. They may be *non-commercial attribution*, which means we can share and adapt the sound, but we must give credit to the original source, and we can't use it in commercial products without contacting the creator and getting permission.

The important thing to know about copyright is that unless a sound is in the public domain, we shouldn't use it in a commercial project unless we've got a written license

Table 2.2 Example spreadsheet to track use of sound effects libraries

Filename	Source	License	Notes
acceleration.wav	https://freesound.org/s/179940/	Attribution	
footsteps.wav	https://freesound.org/s/460083/	Attribution non-commercial	

from the creator/owner of the sound. If you are using sounds for a school project, it's not considered a commercial project, but you should still *cite the source*, even if it is in the public domain, to indicate clearly what is your own work and what you did not record. For these reasons, it's very important when you download sounds from a sound library that you keep track of where you got the sound from and what type of copyright is involved. Having a spreadsheet handy while you work on audio projects is a useful way to keep track (table 2.2).

Exercise 2.17   Find a Sound Like . . .

Find a sound in a sound library that best illustrates the following onomatopoeia: "oof," "pop," "bang," "whoosh," "swish," "glug."

Exercise 2.18   Opposites Attract?

Find two sounds that appear to be opposites or contrast with each other. You can do this with a partner, where one person finds a sound and the other finds its opposite, or on your own: freesound has a "sound of the day" that you can use as a pretend partner.

Exercise 2.19   Exquisite Corpse (Group Game)

Exquisite corpse is a group art/writing game in which every person adds to a composition by seeing only what the last person drew or wrote (not the entire composition). It was invented by members of the Surrealist art movement. Try to create an auditory exquisite corpse with your group or class: create a soundscape, where one person chooses a sound effect, and plays it for the next person, who chooses an effect that will go with the first. The second person then plays only their effect to the third person, who chooses an effect that will go with that, and so on. Have one person sequence the sounds into one long composition (you could do this remotely by having each person email the sound privately to the selected sequencer). Where did you start and where did you end, and what were the connections between the sounds?

2.10 Segues

We're going to be making more sound effects compositions in the coming chapters. We'll look at mixing in the future, but it's worth looking quickly at how we might move from one sound to another. Layering sounds one after the other can tend to sound a bit rough, and if there are more than two sounds it can quickly get messy. We can put them one after the other, but that can lead to quite abrupt jumps between them (hard cuts) unless we wait until the first sound has decayed before adding the second.

If we fade one sound out, and then fade another in, this is called a *v-fade* (figure 2.22). If we fade one sound out while we fade another in at the same time, so there is a place where we can hear both sounds, this is referred to as a *cross-fade* (figure 2.23). Finally, if we fade one sound out and bring in another sound at full volume, this is called a *waterfall* (figure 2.24).

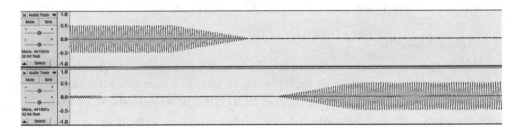

Figure 2.22
V-fade: one sound fades out, then another fades in.

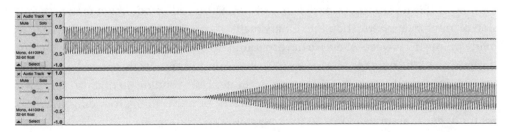

Figure 2.23
Cross-fade: one sound fades out while another fades in.

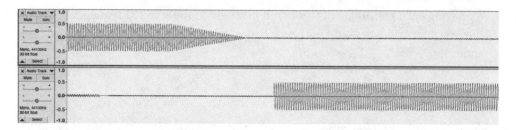

Figure 2.24
Waterfall segue: one sound fades out and another is brought in at full volume.

Exercise 2.20   Segues

Take two sounds and apply different fades to place the sounds together one after the other along the timeline. How does this change the way you hear those sounds?

Exercise 2.21   Loop de Loop

Cut or segue a sound so that its end will match its beginning, and repeat it so that it loops. How does that impact your reception of the sound?

2.11   Digital Sounds

Sound is a form of energy, which means it can be changed from one form to another. For example, sound energy can be turned into electrical energy (the basis of *analog* sound). Sound can also be converted into a digital signal, such as the sounds that we hear on our computer, our iPod, and so on. Imagine analog sound as a hill: it's a gradual slope up and down. *Digital* sound is like stairs cut into that hill: we're on one stair or the next, but never in between. This common method of digitally transforming a sound is called *pulse code modulation,* or PCM. Pulse code modulation is the conversion of amplitude information into binary digits. If we have 20 stairs up a hill, there are many points in between where we have to round up or down to a stair (figure 2.25). But if we increase the number of stairs to 200, there are fewer points that must be rounded up or down, and the slope becomes more accurate. The rounding up or down is known as *quantization*. Quantizing can introduce artifacts, or noise, in the rounding errors, so we try to get as accurate as we can in those stairs.

The process of turning a sound into a digital signal is called *sampling*. Sampling takes a "sample" of a waveform's amplitudes at regular intervals. That sample contains the

Figure 2.25
Analog versus digital sound.

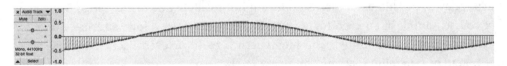

Figure 2.26
Sample rate of a sine wave in Audacity at 44.1 KHz. Each dot at the end of a line is a representation of a sample.

information of the amplitude value of a waveform measured over a period of time. We can think of it as having taken a photo snapshot of the waveform at very short intervals. The *sample rate* is the number of times the original sound is sampled per second, or the number of measurements per second (sample rate is also known as *sample frequency*). A CD quality sample rate of 44.1 KHz means that 44,100 samples per second were recorded. If we zoom into a sine wave generated in Audacity, we can see these stairs as points on the curve (figure 2.26).

The default setting in Audacity is 44.1 KHz. If we change the sample rate ("project rate") to 8,000 Hz (Tracks > Resample), you can zoom in and see that while we still have a curve, it's made up of much fewer points—the number of samples per second has dropped considerably (figure 2.27).

Figure 2.27
Sample in Audacity at a rate of 8,000 Hz.

When the sample rate is more than twice the highest frequency being sampled, we can faithfully reproduce the original sound: in other words, if we have a sound that peaks at 4,000 Hz, a sample rate of 8,000 Hz will be sufficient. This is known as the *Nyquist principle*. If lower sampling rates are used, then we lose some of the original or we get artifacts known as *aliasing*. Given that human hearing tops out at 20 kHz, we commonly sample at 44.1 kHz. But wait, that's not twice 20! Why do we leave some additional room for sounds if people can't hear them? Well, we know there are sounds above our threshold that may be harmonic components of a sound: it's quite possible we can feel or sense these frequencies, even if we can't hear them. These ultrasonic sounds may have an impact on how we hear sound. Some people claim that a higher sample rate sounds better to them. It may not, but given that we are not usually limited today in our file sizes, there's no reason not to allow this higher range. We could, of course, sample at an even higher rate, and if your device allows and space on your memory card is not limited, you might record at 48 or even 96 KHz sample rates, although bear in mind that any processing on these sounds later takes a little bit longer as your computer works harder to crunch the numbers.

Bit depth is used to describe the number of bits (amount of data) recorded for the sample, and you can think of it as the amplitude equivalent of sample rate. The bit depth corresponds to the *resolution* of the sample: as with sample rates, higher bit depths result in higher resolution, better quality or *fidelity*, but larger file sizes. Variations in bit depth affect the quantization error rate and dynamic range. Higher bit depths bring an increased *dynamic range* of the sound: the distance between the highs and lows in amplitude. A lower depth will "crush" the sound into a lower dynamic range. A 24-bit digital audio track has a dynamic range of 144 dB, compared to 96 dB for 16-bit (in theory, but in practice it's usually lower). CD quality sound is considered 16-bit, although often the CDs are recorded in 24-bit and converted to 16-bit before release. An 8-bit sound is considered land-line telephone quality in terms of dynamic range (note that telephone quality also restricts frequency, which is the most obvious difference you hear). Bit depth is important to audio only if we are using uncompressed pulse-code-modulation sound (PCM and variants like ADPCM and LPCM). To resample

Table 2.3 Commonly used sample rates, bit rates, and bit depths

	CD audio	DVD audio
Sample rate	44.1 KHz	192 KHz
Bit rate	ca. 1.5 Mbps	5 Mbps
Bit depth	16-bit	24-bit

in 8-bit, we need to export our file and bring it back into our project (File > Export > Export as Wav > Save as Type > Other Uncompressed Files > Header: WAV > Encoding: Unsigned 8-bit PCM). The file may look no different in Audacity, but if you open the spectral analysis in a program that provides a more detailed spectrogram, like Adobe Audition, you'll see a difference—the increased dynamic range allows for more light and dark variation.

Finally, we have the *bit rate*, which is the number of bits per second that can be encoded and decoded (processed/played back), represented as kilobits per second, or kbps. Higher bit rates are higher-quality sounds, just like sample frequency rates. When we save a sound, we can usually save it using a constant or variable bit rate. If you choose variable, the compression technology makes the decision as to where it should apply more of a change in bit rate. Acceptable bit rates for audio are usually a minimum of 128 kbps. A sound at 128 kbps is equivalent to sound you'd hear on an FM radio. It is unlikely you'll notice much quality improvement above 216 kbps unless you have high-end equipment. For reference, Spotify currently uses a normal bitrate of 160 kbps on desktops and 96 kbps on mobile phones, and if you have a paid membership, you *may* have rates of up to 320 kbps (Spotify's Hi-Fi option). What is happening in the compression is that an algorithm removes bits from the track based on our understanding of human hearing, assuming you won't hear the sounds, thereby reducing the information processed per second.

2.11.1 Lossless and Lossy Formats

When there is no loss of data in sound file conversion, we call it a *lossless* format. Lossless formats include wave (WAV) and FLAC. *Lossy* formats, on the other hand, decide what is important and throw away some data to compress the file size. While compression formats are now very good, the more we compress a file size, the more data gets thrown out, and since what gets thrown out is determined by an algorithm—not our ears—we can sometimes hear the loss in comparison tests. Common lossy formats include MP3, M4A, OGG, and AAC.

Knowing what we now know about sample rate, bit depth, and bit rate, why do we use lower rates to compress files at all? The answer lies in file size: the higher the rates, the larger the file size. In the earlier days of the internet when it would take half an hour to download a song, the MP3 format became very important. The difference in file size between an MP3 and uncompressed wave (WAV) file would be about 4 MB for MP3 and 40MB for the wave file. These days, streaming delivery services mean we still want smaller file sizes. When a client or instructor has asked for a submission as a compressed file, we always work with uncompressed files and then *downsample* the file by compressing it at the last possible stage in our production. This ensures a higher-quality end product. So even though we may be submitting a file as MP3 for a podcast to be streamed, we would probably record in at least 44.1 or 48 KHz and work on it as a wave file until we have finished any manipulation or effects, and *then* finally save it as an MP3.

Exercise 2.22   Compression Fidelity

Grab a sound effect that is uncompressed (a wave file) from a sound library. Adjust the sample rate in stages (`Tracks > Resample`). What do you notice as it approaches 8 kHz? Export the original as MP3 (`File > Export as MP3`—you may need to follow the instructions to install the LAME MP3 codec first). Change bit rate mode to constant and then select the lowest quality (the bit rate) and export. Listen at lower quality. Bring that saved file back into Audacity and listen and compare to the original. Don't forget to look at the spectrogram as well, and turn up the volume!

Reading and Listening Guide

Jonathan Sterne, "The MP3 as Cultural Artifact" (2006)
In this article, Jonathan Sterne explores the design and implications of the MP3 format from an industrial and *psychoacoustic* perspective (the science of how we perceive sounds). Many *audiophiles* (people who love audio, and have a tendency to purchase high-end stereo equipment) have claimed that MP3s are really poor quality. Neil Young even took to Kickstarter in 2014 and successfully raised millions of dollars to launch his Pono uncompressed audio format, arguing that MP3s were inferior. Unfortunately, two years later the format was killed after being purchased by Apple. Are MP3s really inferior? Or is it down to how much compression is used in an MP3?

Jonathan Sterne, "The Death and Life of Digital Audio" (2006)

Sterne tackles the debate between analog and digital recording in this article. Have you wondered why vinyl is seeing such a massive resurgence? Although the article is becoming dated, the debates are still relevant today. Read the article and think about your own opinions about vinyl. Which is better—analog or digital?

Marc Perlman, "Golden Ears and Meter Readers: The Contest for Epistemic Authority in Audiophilia" (2004)

Although this has a very academic title, Perlman tackles the concept of the audiophile in this article in a reasonably accessible way. Perlman explores the truth versus fiction in our notions of audio fidelity. Perlman calls for ABX tests—that a computer plays two sounds, A and B, and a third sound, X, where the listener has to match X to either A or B. Under computer-randomized trials, most people who claimed to be able to hear a difference actually couldn't. Can you? There are many lossless audio tests online. See if you can hear a difference. It helps if you have fresh ears that aren't tired from a day's listening, and an excellent sound system.

There is an awful lot of very expensive audio equipment you can purchase. A lot of audiophiles argue that the more expensive equipment is worth it because the quality is superior. However, the James Randi Educational Foundation set a challenge in 2007 offering a million dollars to anyone who could prove scientifically that the $7000 Pear Anjou speaker cables were actually any better than your everyday, off-the-shelf $100 Monster cables. After nobody could prove the cables were better, Pear cable pulled out of the test (see White 2007 for an overview). It's important to be cautious before shelling out for the expensive equipment; a lot of marketing sounds scientific, but it's not. Do your research. Just because it costs a lot more doesn't mean it's higher quality or it's worth spending your money on. There are some times when the money *is* worth it, but nothing beats a good pair of ears!

3 Recording Sounds

In chapter 2 we discussed how library sounds can be under copyright, and can also become overused and recognized by audiences. When we can, it is always best to use our own recorded sounds rather than those of a sound library. Of course, sometimes sound libraries are necessary, because we don't often have the access, time, or budget to go record the cockpit of a B52 bomber, a Japanese Shinkansen bullet train, or a lion's roar, and so on. But if we are serious about sound design, we should start recording and collecting our own sounds into our own sound library. Part of that comes down to not wanting to duplicate what's already out there, and part of it is about having an original sound that is *ours*, so that we own the copyright and have control over how it gets used. In this chapter, we'll begin exploring recorded sound, looking at some basic aspects of microphones, the associated tools, and recording techniques. Like listening, the more we practice, the better we will get at recording sound, and the more comfortable we will get with the equipment involved.

3.1 Audio Slating

When we record sound, it's important to *slate* our audio. An audio slate is equivalent to the slate used in film production, which you've probably seen: someone scribbles some notes on a clapperboard and yells, "Take one!" and snaps the clapper, also known as a slate. The audio slate isn't on a clapperboard but is just us speaking some notes into the microphone at the start of a recording, and is designed to give us some basic information: time of day, and date; temperature and wind conditions (if outside); location; subject being recorded; recording gear used. In other words, any information we may need later to identify or share the file. Without slating, we might forget which of the sounds was which bird or which car engine we recorded, or what the sound itself was. Going back to files months or years later, having the information about the sound file can be really helpful.

Be sure to also leave two seconds at the start and end of the recording; you can cut this down later if needed, but it gives you some room for fade-ins and outs. If we're recording with a video camera or multiple microphones, it helps to slate the multiple recording devices with a hand clap if we're not using a clapperboard, to help with synchronization across devices later, although auto-sync is getting much better. If we can't slate the sound at the beginning because we whipped out the recorder at the last second to catch a fleeting moment, then slate at the end of the file before hitting "stop"—this is called *tail slating*. At the tail end, we might also want to note things that happened during the recording (for instance, "airplane flew overhead half way into that, rendering it useless. Delete this file"). These notes help us quickly navigate useful or useless files when we come to organize our files later.

For example, we might slate a field recording like this:

> February 14, 2019, 9:45 a.m. Clear, sunny day, with snow and ice on the ground. Backyard of my house recording the cheerful male cardinal that comes to sing in my blue spruce tree, approximately 10 m from my house where I'm standing inside the doorway out of the wind. Using a Sennheiser MKH-416 shotgun mic on the H4N recorder.

Imagine if we didn't know what a cardinal sounded like, but knew what one looked like, and we were collecting a lot of bird sounds that day. Going back six months later and trying to figure out which one of thirty files recorded that day was the cardinal sound might be useless. Likewise, what if we're recording a bunch of sounds with several different microphones, and we really loved one, but couldn't remember which mic we used to go record similar sounds with? We'd have to repeat the whole recording again. Slating also helps with naming the files later, with cataloging the files in our own library, and with sharing them with others. It may seem like a fiddle at first, but your future self will thank you.

3.2   Stereo or Mono Recording?

You've probably come across the terms *mono* and *stereo* on music recordings (or "monaural"/"monophonic" and "stereophonic"). Back in the early twentieth century, mono files were the norm, because the earlier record players had a single speaker. Mono sound was originally designed to play the sound through that one speaker, and records until the 1950s were recorded in mono. By the late 1950s and early 1960s, recording technology and home consumer audio products had advanced, and people started using two *channels*, mapped to two speakers to play sound—stereo. Each speaker would have a different channel—a different track played to that one speaker. Bands started to play with stereo techniques, which enabled them to create a sense of space in the recording by panning a sound to one side (adjusting the recording so that, for instance, the lead guitar would be heard in the left speaker, while the bass guitar might be in the right). Stereo came to be associated with "quality," and mono was phased out by the late 1960s.

Many people believe "more is better" in most aspects of their lives, and while stereo enhances the playback of many types of sounds, when it comes to *recording* sound, stereo is not always better than mono. Unless we are recording a wide space, recording with stereo microphones, or the sound is moving, stereo is often unnecessary, since we will likely have the same sound in both the left and right channels. In that case, we are just doubling the file size by duplicating the same signal, taking up more room on our recorder's flash card than necessary. We are effectively creating not stereo, but *dual-mono*. Most one-shot effects in sound libraries, and most vocals for dialogue are recorded in mono because there is no need for stereo. For many purposes, if we need a stereo file, we can resample our file later to split the mono signal into two and digitally alter the stereo position (we'll come back to this later). On the other hand, recording ambience—where we need a sense of space—or recording something that is moving is often done in stereo. For this reason, a recording device will give us the choice to switch between mono and stereo. Stereo recording requires two microphones, or a specially designed stereo mic, since we need to capture two separate channels; most of the recording we'll focus on will be mono.

3.3 Microphones and Microphone Selection

There are many different types of microphones, and they range in cost from a couple of dollars to many thousands of dollars. What difference does the microphone make? Quite a lot! Each type of mic has a different purpose, so we can't really say which type of mic is "better" than another, only which is better *for a specific purpose*. There is so much to microphone selection and use that many books have been written about it. We will just cover the basics.

The first thing to understand is that microphones amplify sound in different ways in terms of the technology they use. A microphone is a transducer—it converts energy from one form to another, and the technology used to do that conversion determines the type of microphone it is. The most common microphones we encounter are probably dynamic or condenser microphones.

3.3.1 Dynamic Microphones

Dynamic microphones have a diaphragm with a long coil of wire wrapped around a magnet. Sound pressure moves the diaphragm, which moves the coil, causing an electrical current to flow. Dynamic microphones are quite rugged (they can take a bit of beating and be dropped on stage, for instance), usually comparatively affordable, and don't require batteries or external power. They also tend to have a great signal-to-noise ratio. We can usually tell if a microphone needs external power by the presence of an on/off switch on the mic: if it has a switch, it usually doesn't need external power. Dynamic mics can also be fairly resistant to environmental changes like temperature or humidity.

If we take a look at the frequency response charts of dynamic mics, though, we notice that they often have a boost in one specific range, and therefore are usually selected for a specific purpose based on this response. Dynamic mics are best for low- to mid-frequency sounds like drums, electric guitar, or vocals. We can also buy specialized kick drum mics, or bass mics, designed for low-frequency sounds. A popular example of a dynamic mic is the Shure SM57, which has a great response for vocals, and is very rugged (you may have seen artists throw them on stage, pick them up again and keep singing). They're much better for recording very loud sounds (like gunshots and drums), as well as recording outdoors, because they're pretty solid.

3.3.2 Condenser Microphones

The second most common microphone type used in professional recording is the *condenser mic*. These use a lightweight membrane with a capacitor on either side of it. Sound pressure on the membrane fluctuates the capacitor, creating an electrical signal. Condenser mics have a more uniform frequency response than dynamic mics, particularly in the high and low frequencies. They are great for getting detail in a sound, but are more delicate—a loud sound too close to the mic can blow the membrane and destroy the microphone. They are also prone to problems in high temperatures or high humidity, and they are generally more expensive. They may require external power sources, known as *phantom power*, which amplifies the weak voltage they output (more and more modern condensers have built-in power now, so you'll have to figure out what you need by listening or looking at the manual). Condenser mics, because they can capture high-frequency sounds, work better on instruments that have high frequencies or lots of harmonics in higher frequencies, such as cymbals or acoustic guitar. The size of the diaphragm influences the frequency response. A large diaphragm is used for voice, for instance, while smaller diaphragms get a better response on the higher frequencies. A condenser microphone that is popular with professionals is the

Neumann U87. It costs several thousand dollars, but you get what you pay for: the frequency response is nearly flat across the entire spectrum, making for a highly accurate recording.

Condenser and dynamic mics have different *transient responses*. Transient response refers to how quickly the mic responds to changes in the sound. Dynamic mics don't respond as quickly, so their transient response isn't as accurate as a condenser mic. Most people who work in sound gather a collection of microphones for recordings, because they each have their own unique characteristics. Small-diaphragm condenser microphones are an easy default while you get used to microphones, and are probably the most versatile mics. The mics on most handheld recorders are small-diaphragm condensers, because they are the most useful for a variety of types of recording.

3.3.3   Other Types of Microphones

The vast majority of microphones you will encounter in professional recording environments will be dynamic or condenser mics. We'll be building and exploring other types of microphones below, but some other types of mics include:

Ribbon microphones: A thin strip or ribbon of aluminum or film is moved between two magnets by the vibration of sound. These are very delicate and very expensive, but modern technology is improving their durability and lowering their cost.

USB microphones: USB is becoming a more commonplace way to use microphones for podcasts or if you're on a budget. Most USB microphones are small-diaphragm condenser microphones, like the Blue Yeti. They are designed to plug into a computer and record digitally, rather than for use with professional recorders.

Stereo microphones: Stereo microphones are dual microphones, designed to pick up two signals from the same sound at once from different directions—usually in what is known as an X/Y position. The angle between the microphones typically varies between about 90 to 135 degrees, with a wider angle corresponding to a wider stereo image.

Binaural microphones: These are dual mics designed to be worn on or in the ears. We'll get into binaural sound in chapter 7, but for now, you're unlikely to encounter them except in specialist circumstances. They are usually condenser mics.

Surround microphones: With new technologies, surround sound microphones have become available, such as Holophones, or the DPA d:mension 5100, a 5.1 surround mic (shaped like and called by some "the bicycle seat"). These are large capsules with an array of microphones inside the capsule that capture each direction. In this way, they're designed to capture five separate sounds into five different microphones on location. Other approaches include double-mic approaches that use processors to segregate the signal. They are usually condenser mics.

Figure 3.1
Stereo microphone: the RØDE NT4.

Lavalier microphones: Lavalier ("lav") and lapel mics are designed to be clipped onto the clothing—you will often see these on people being interviewed on television. While they are fine for voice on sets, they're not usually used for other sounds, although they're handy for some tasks where you need a small microphone.

Contact microphones: Contact mics consist of a piezoelectric disc that must be placed directly onto a vibrating surface (i.e. in contact with it). These can record some really interesting sounds. Hydrophones, underwater microphones, are usually built with piezo discs as well, with the addition of a transducer and some form of a container to protect the wiring from the water.

3.3.4   Polar Patterns

In addition to responding to frequencies differently, microphones also have *directional* properties. We'll be dealing in more depth with directionality and spatial audio later, but for now, a basic understanding of how mics pick up sound is useful. The directionality of a microphone is described using what are known as *polar patterns*, or sometimes pick-up patterns, because they illustrate where the mic picks up sound. Polar patterns describe the directionality of a microphone on a sphere, but they're often visualized in a two-dimensional circle.

Omnidirectional (or just *omni*, as in "all") microphones pick up sound in all directions evenly. Assuming the microphone is at 0 degrees, an omni microphone's polar pattern shows similar sensitivity in the 360° range. This means an omni mic isn't aimed at a particular sound source—it will pick up sound all around it regardless. At higher frequencies omni mics can become directional, meaning that the sound will need to be in front and *on axis* (see below) to be picked up, so high-frequency sound arriving from behind the mic can sound a bit muddied or dull.

Unidirectional microphones will pick up most sound in one direction—for this reason, they're usually just called *directional mics*. They may pick up some sound from other directions, but they are more sensitive to one particular direction. Cardioid (heart-shaped) patterns are the most common—with most sensitivity on axis (0°) and least to the sides, at 180°, but with a spread of about 130°. Unidirectional mics will pick up less of the ambient sound than omni mics. *Shotgun* microphones are commonly used in film and field recording to pick up a specific sound, as they are very unidirectional microphones. Even microphones that are called unidirectional are not completely unidirectional, though, and they will pick up some sound behind and/or to the sides. Shotgun mics are useful when we need to focus on one sound and filter out a lot of ambience—while an omni or cardioid mic might be useful, for instance, in picking up the ambience of a forest, a shotgun is useful for picking out a specific bird in that one tree, and focusing on that.

Bidirectional mics usually pick up in front and back, with less coverage at the sides. Supercardioid and hypercardioid patterns for instance will pick up sound in front about 115° for supercardioid and about 105° for hypercardioid, with varying degrees in the rear.

It's important to remember that the polar patterns are described as they record in optimal conditions, and are determined in anechoic chambers (special rooms that have no reflections). Putting a microphone near a wall or floor will change the sound that gets picked up (we'll talk about the environment and reflections in chapter 4). Some mics allow you to change the polar pattern at recording time. Others even allow you to change it afterward using software such as Sennheiser's Ambeo Pattern for the MKH 800 Twin mic.

Exercise 3.1 Microphone Selection

If you have access to multiple microphones, try recording the same sound with different microphones to hear how they pick the sound up differently. What differences can you hear? Look up the microphone's frequency response. Did the differences in what you hear align with the frequency response charts? Don't forget to slate your audio!

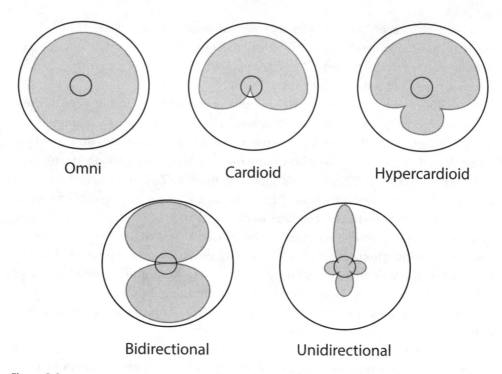

Omni   Cardioid   Hypercardioid

Bidirectional   Unidirectional

Figure 3.2
Polar patterns: these are spherical, not two-dimensional patterns, so an omnidirectional microphone records in 360° in a spherical shape from the tip of the microphone.

Exercise 3.2   Build a Contact Mic

A contact microphone picks up vibrations through *contact* with objects—sounds transferred through solids, rather than air. A *piezo* is a metal disk with ceramic piezoelectric crystals in the middle. The vibration produces a buzzer sound, so you'll find piezo speakers in all kinds of toys that make sound that you can hack.

Equipment

a piezo disc, sometimes called a "buzzer," of about 27 to 30 mm

about 30cm of balanced microphone cable, e.g., Mogami W2582, or ¼" guitar cable

a ¼" in-line female mono jack (TS)

wire strippers/cutters, solder, and soldering iron

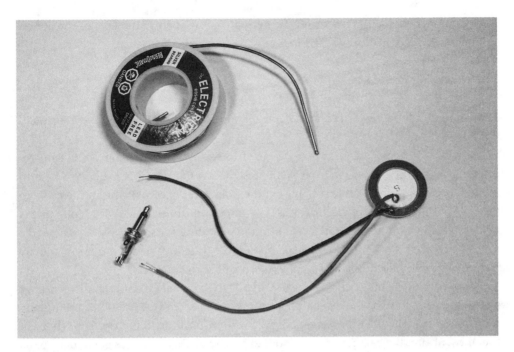

Figure 3.3
A contact microphone.

The piezo has two terminals: the inner terminal (the signal) and the outer metal disk (the ground), so we need to use a cable that has two conductors (e.g,. a microphone cable). Strip about 5cm off the end of the cable, which should reveal the two conductor wires inside. Separate the copper shield wires and the conductor wires (blue or red and clear). Twist the two separate groups of wires into separate bundles.

The sound signal will be weak, and so you'll get a lot more high-frequency sound. Although you can record louder sounds with a contact mic, to get more delicate sounds and a balance of lower frequencies, you will need to purchase an amplifier/preamp.

Exercise 3.3   Use an Inductor Coil to Record Hidden Sound

Inductor coils convert electromagnetic radio waves into sound. We can quickly and easily build an inductor coil, or quite cheaply purchase them online as telephone pickets or "circuit sniffers." These electromagnetic waves can be found in any electronic devices, from lights to remote controls. Inductor coils can even pick up electromagnetic radio waves from far away in space, such as meteors or the aurora borealis. The more powerful the inductor, the more they pick up waves.

3.4   Recording Accessories

Many microphone accessories are available for purchase to improve our recording quality. These include:

Boom pole: A boom pole is a pole designed to be held by the hands with a microphone at one end. Some boom poles are designed with the cable inside the pole to reduce the sound of the cable rubbing on the pole, but we can also strap the cable to the pole with tape, clips or elastic bands. The important thing to look for is one that is lightweight but sturdy, and you feel comfortable holding for some time. Some people prefer to wear gloves while holding the boom pole to reduce hand noise, so you may want to try out some gloves and see if you can still keep your grip.

Microphone stand: Mic stands are really useful when you are recording solo, since we need our hands free, so holding a boom often isn't practical. Mic stands are usually some variety of tripod with a longer pole that can be positioned on top. The weight of the mic will alter the balance of the tripod, so be sure to check that it's got a stable base, and is not going to drop the mic onto the ground once the mic is attached.

Shock mounts (figure 3.4): Shock mounts are designed to reduce handling noise and mechanical noise of microphone stands, and they usually suspend the microphone using elastic cords, and attach to the end of a boom pole or mic stand. Shock mounts come in a lot of shapes and varieties, for indoor or outdoor use. Most mics will come with their own fitted holder to attach to a microphone stand (and some will come with their own shock mount built into the holders). There are differences between American and European threading sizes, so you will need an adaptor to attach some of these to different-sized stands.

Windscreens: These come in many different types, from basic pop filters designed for indoor use to reduce popping on the plosives in speech, to windjammers, "dead cats"/ wombats, "blimpies," "zeppelins," and "furries" put over the microphone unit for use in outdoor windy environments. Most microphones will come with a foam windscreen for very basic reduction of environmental noise. It's good to have a collection as they have advantages and disadvantages in different environments.

Pads: You can purchase pads that will reduce the decibels picked up by a microphone in order to record in very loud spaces, or with very loud sounds. They fit on the end of the mic before the cable, to reduce the signal going into the cable.

Figure 3.4
Shock mounts.

Phantom power: As described above, some microphones require a phantom power setup. Some of these are battery operated, and some are built into professional recorders. It's important to know if your mic needs power. Phantom power units can be attached to mics before they hit your recorder, but most recorders now have a built-in phantom power.

Preamps: Sometimes just called a "pre," the preamp converts the weak output of some microphones into louder signals for recording. As with phantom power, you can buy external units, but many professional recording devices have built-in preamps that can be switched on or off.

Pop filter (figure 3.5): These are used for vocals. The screen is a simple barrier that scatters the puff of air caused by the plosives in our speech ("P" and "B" in particular). The pop screen should be about 10 cm from the mic.

Plugs and cables: Most professional mics have XLR plugs. We may need adaptors if our input takes ¼" TS (tip sleeve) or TRS (tip ring sleeve), balanced or unbalanced phone cables, like headphone jacks. Others have USB adaptors. Knowing what our microphone cables are and what adaptors we need to connect to our recording device is important. The more adaptors we have to add, the more chance of there being

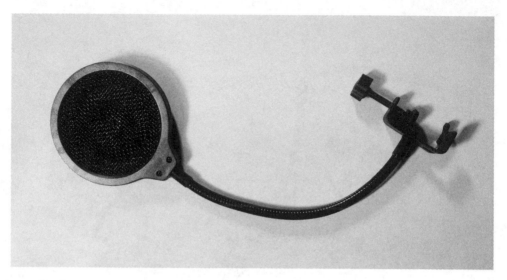

Figure 3.5
Pop filter.

interference or problems with the recording. Different types of line inputs have different requirements in terms of sensitivity. An external sound card, for inst ance, might have TS *line inputs*, which are usually for stereo recording and tend to require a signal boost of a preamp unless they have a built in preamp. Balanced cables have signals run along three wires: positive, negative, and ground. Balanced cables eliminate noise with noise cancellation. XLR and TRS cables are balanced. Unbalanced cables can add noise to a signal, so should be avoided for recording unless you are recording in very short distances (5 or 6 meters).

Recorder: Portable recorders are commonly used inside and outside the studio. Zoom and Tascam make some very popular, affordable recorders. Many will have built-in phantom power and preamps, can input XLR or ¼" balanced or unbalanced cables, 1/8" (3.5 mm) mic/line mini phone jack, mic and line-level signals (no external preamp required). Look for one that also has built-in mics you can use in a pinch. It's also possible to purchase XLR adaptors for a smartphone and record onto the phone from a standard mic, and there are some mics designed with smartphone adaptors that can now fit onto the phone.

It's best to purchase a dedicated recorder if you can afford one, because it will have all of the additional settings you may need, like phantom power and preamps, but a phone adaptor is handy for everyday travel purposes. Some expensive recorders also

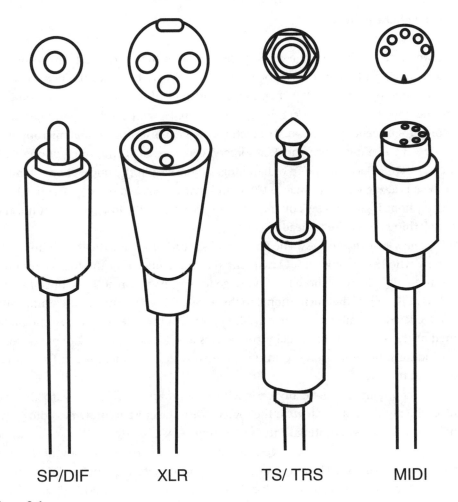

SP/DIF   XLR   TS/ TRS   MIDI

Figure 3.6
Common audio cables.

exist—like the Sound Devices 702. The Sound Devices is usually up and running within just a couple of seconds of it being turned on, whereas (in my experience) other pocket recorders the Zoom can sometimes take over two minutes to boot up and be ready. In addition, there are settings like time codes for film that can be used on the more expensive equipment, but for basic sound design purposes a portable recorder like the Zoom H4N is a perfect balance of affordability and capability. Be sure to check if you also need to purchase a memory card with your device.

3.5   Microphone Position

Where we place the microphone can significantly alter the sound that gets recorded. Of course, paying attention to the polar pattern is going to be relevant when we use a directional microphone—in that case, the axis or angle of the microphone is going to determine how much of the direct sound we pick up. How we hold the microphone is also going to influence the sound. As described above, most professional microphones are designed to be used with shock absorbers, which suspend the microphone with elastic cables to reduce the sound of handling. Where we put the microphone stand or boom pole is also going to influence the sound that gets picked up (e.g., the height of the microphone is going to pick up more footstep sounds if it's close to the ground, and more breathing if it's close to head height).

Side-address versus top-address: The position of the capsule inside the microphone will determine whether the mic should be pointed at the sound from the top or from the side. Hand-held mics (like those used on stage for voice) are top-address, meaning the capsule is at the top of the microphone, so the mic should be pointed at the sound with the top. Side-address mics have the capsules pointed at the side, so the side should be pointed at the sound source. Usually condensers are side-address, and dynamic mics are top-address, but not always. Consult the manual for the microphone to discover more about the mic.

Axis: Microphones are generally designed to record on-axis (that is, with the end pointed directly at a 0° angle toward the source), with the most accurate recording having the correct address pointing at the sound source. As the angle of the microphone toward or away from a sound changes, the frequency response of the sound that the mic picks up will change. The amplitude will drop, and high frequencies and harmonics will tend to get lost as you move off-axis, leading to a slightly "muddier" sound. On the other hand, if we're recording certain sounds, like voice, being off-axis can reduce the pops, pick up more "room tone," and lead to a warmer sound (figure 3.7).

Exercise 3.4   Off-Axis Recording

Try recording the same sound with the microphone positioned at different axes. What changes do you hear in the frequencies that get picked up and the amount of environmental sound?

Distance: The distance from the microphone to the sound source also influences the sound. The invention of the electric microphone led to a new style of singing in

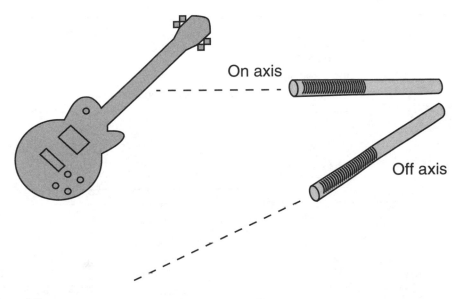

Figure 3.7
On- and off-axis recording.

the 1940s, called crooning, exemplified by Billie Holiday, Bing Crosby, and others. The fact that the microphone could pick up more delicate sounds meant that singers could be much softer than previous recordings, and the mic picked up nonvocal emotional effects like breath, the tongue clicking on the teeth, and other sounds that previously couldn't be heard in music recordings. These delicate sounds place us in a very intimate position relative to the singer, since to hear those types of sounds in "real life," we'd have to be standing very close to the vocalist. Physical distance, then, can create a perceptual distance (we'll come back to this below).

Ambient sound will be picked up by any microphone. How much ambience we want should be a factor in our selection of microphone, in addition to how close we place the microphone to the sound source. Directional microphones can be placed farther away from sound sources because they won't pick up as much ambience as omni or bidirectional mics. An omni mic will need to be about twice as near to a sound source to reduce ambience. As a microphone is moved closer to the source, more of the direct source and less of the environmental sounds get picked up.

Proximity effect: As a microphone is moved closer to a source, the bass response will increase. If we place a directional microphone close to a source (i.e., about 30 cm), we will get a *proximity effect* of having emphasized low frequency (< 200 Hz) content.

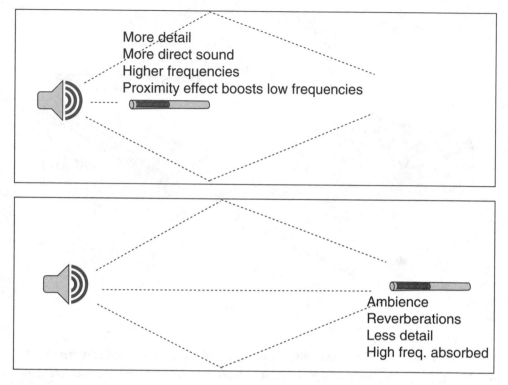

Figure 3.8
The distance of a microphone from the sound source changes what sounds get picked up.

This can make a man's voice sound rich and deep, but can impact speech intelligibility and can make low frequency content in nonvocal sounds sound a bit muddy, and can make some instruments (e.g., acoustic guitar) sound too "boomy." The proximity effect is most apparent on a bidirectional microphone, and least (or not at all) on omni microphones.

Exercise 3.5 Proximity Effect

Record sounds at close proximity and then back off to a distance of more than 30 cm without adjusting the input gain. Can you hear a difference? What changes as the sound gets closer to the mic?

Feedback: Feedback loops are created when the microphone is too close to an amplifier/speaker monitor, or turned up too loudly, resulting in an increasing loop of loudness that gets amplified, creating a high-frequency ringing sound. Move away from the amplifier (and point the microphone away from the amp/monitor), move closer to the sound source, or turn down the microphone's volume (or input gain—see below) to reduce feedback. Flatter response mics are less likely to cause feedback, and non-omni mics will reduce feedback noise. Feedback can be a wanted effect (the Beatles, for instance, probably used it first on "I Feel Fine" in 1964, and other bands have used it since), but usually it's not wanted at recording time.

Phasing: When we are using two or more microphones, we can run into phasing problems. When the same sound arrives at two more mics at different times, because of the difference in arrival time, a destructive interference pattern can occur, causing a "whooshing" sound. You may have noticed this in a classroom if the instructor was wearing a lapel microphone and had the lectern microphone also turned on. As they move around the room, phasing effects may be picked up, as the time it takes for their voice to reach the amplifier varies between the two microphones, shifting one signal out of phase with the other. The destructive interference we discovered in chapter 2 can happen between just a few frequencies, and we may not notice at recording, but we may notice later that we're missing out some frequencies because they've been canceled out by phasing issues (figure 3.9).

Phasing can also occur if we have a single microphone close to a reflective surface. The reflection and the direct sound occur at slightly different times, resulting in a phasing effect. If we need to put a microphone close to a surface, putting something soft on the surface will cut the reflections. We'll take a look at intentional phasing in chapter 4, but usually it's best not to introduce phasing at recording even if it is a desired effect.

Input gain: The gain is the volume of the input of the microphone. If we plug in headphones, gain is not the same as the *volume* we hear in our headphones: The volume knob adjusts the amplitude of what we hear, and the gain knob adjusts the amplitude of what is being recorded. We need to look at the meters on our recording device to determine the input volume. We typically want to record where the meters are all green just touching to the edge of the (usually yellow) warning (this may also be red).

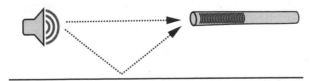

Figure 3.9
Phasing created by reflection onto a microphone.

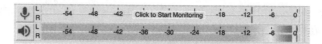

Figure 3.10
Level meters in Audacity.

If it goes into the red it's peaking, and the sound will clip. When we record at too high or "hot" an amplitude, the sound will distort and become "crunchy" sounding. This is known as clipping, as we learned in the previous chapter, and is very difficult to recover from. It's important when we record not to clip the audio. Even if we want a distorted sound, it's better not to record it clipped but instead to add distortion effects later onto a clean sound, so we can determine exactly how much distortion we want.

Volume meters are handy to track how hot the gain is. In Audacity, we can *monitor* before we start recording by right-clicking on the record level meter and then selecting "start monitoring" (figure 3.10). You'll see a red-yellow-green bar appear. The right side of the bar is the *current peak* of the amplitude. You'll also see a line that appears (which can be red, yellow, or green depending on the gain), which is the *recent peak*. Beyond that is the maximum peak, a blue bar that sets the maximum peak in that channel during the session—how high you've gone since starting. There is also a clipping indicator, a red bar to the right of the maximum peak that is highlighted if we have clipped part of our sample.

What is the best volume to record a sound? It can depend on what we're recording, but generally, stay in the green: many people believe that a peak at –20 dB is the optimum recording level. As long as we know we're not risking clipping, we can increase that. If we want an intimate sound but getting too close will cause the signal to clip even at a low gain, we can put a pad on the microphone to reduce the input gain into the recorder.

Exercise 3.6   Input Gain

Make some different sounds with a variety of objects at different distances, and see if you can guess what the best input gain is before you monitor the sound to check.

3.5.1   Stereo Recording

As described earlier, we tend to record spot sounds in mono, and ambience or sounds with some width or movement in stereo. Stereo recording involves two microphones: each mono signal from each mic is assigned to a channel (left or right) in the stereo

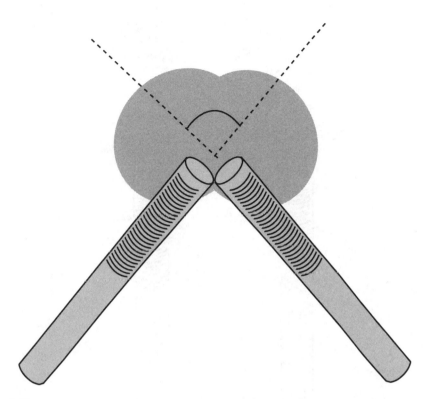

Figure 3.11
Stereo X/Y position of microphones.

file. We want to use stereo if we want a difference between the channels: a difference in timing or a difference in frequency. We want a time difference, for instance, if we want to record a helicopter flying from our right to the left. We may also want a difference in frequency: as we saw, changing the axis will change the frequencies that get picked up, so recording a sound with two different axes means we can bring out different aspects of that sound.

How we position the mics depends not only on what we want to pick up, but also on the polar pattern of the mic. I mentioned the X/Y position in discussing stereo mics above, but if we are using omni mics, we might want to position them in A/B position: side by side, at a width we decide to be as wide as we want (figure 3.12). The trouble with A/B positioning is that we can run into phase cancelation problems from the difference in timing between the two mics. The X/Y position uses directional microphones, and reduces phase cancelation, since the mics are positioned at the same point so that the sound arrives at the same time on both mics. Sometimes, if the sounds are

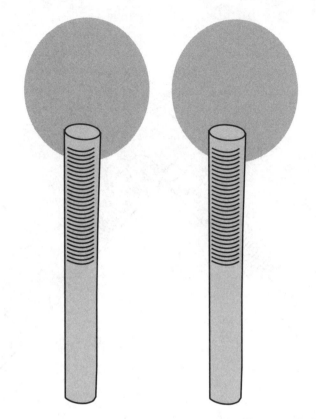

Figure 3.12
Stereo A/B position of microphones.

mixed down to mono, we can run into phase cancelation, but it's rare that we'd want to record in stereo and mix to mono.

Mid/side (or M/S) uses (usually) one cardioid or omni mic, and one bidirectional (figure 3.13) mic. The bidirectional mic is placed at a 90° angle from the source to record the "side," and the other mic points at the source, and functions as the "mid" (figure 3.13). The stereo width can be adjusted to increase or decrease the spread between the two channels, allowing us more freedom than X/Y. The difficulty with mid/side stereo recording is how to mix the signals together—they need to be *matrixed* and decoded (the side signal is split into two separate channels and then merged again). Because of this difficulty, some recorders, like the Zoom H4N, have a built-in M/S decoder to make the process much easier.

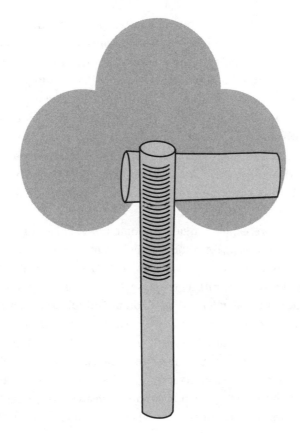

Figure 3.13
Stereo M/S position of microphones.

3.5.2 Proxemics

As we saw above, recording distance is picked up by our brains and affects our perceptual distance to the sound. It's possible to think of the distancing effect of microphones in terms of *proxemics*. Proxemics is the social construction of distance between people: we all have a distance that feels intimate, personal, social, or public depending on how well we know people. The more comfortable we are someone, the closer they can physically be to us without us feeling uncomfortable. Most of us have heard someone complain about someone invading their "personal space": we can also invade their personal space perceptually with sound!

The *intimate zone* is, as the name suggests, for intimate contact or whispering: it's from 0 to 46 cm (18 inches) from us. The *personal zone* is for people we're comfortable around, but not intimate with (friends and family), and ranges from the intimate zone (46 cm) out to about 120 cm (4 feet): but this distance also depends on context. We may be fine with someone being a meter away from us in the queue at the grocery store, but when more room is available, as when we're standing around outside, if someone enters that zone it may feel uncomfortable. We fully expect people to be in our personal or even intimate zone on a packed subway train, although most of us feel uncomfortable with strangers that close to us. The *social zone* is for people we know socially, but more on an acquaintance level, and ranges from the personal zone up to 3.7 meters (12 feet). Finally, there is the *public zone*, which is farther than about 3.7 meters from us (figure 3.14). These proxemic zones are not hard-and-fast rules, and vary by culture.

In film and television, camera angles are sometimes spoken of and used in terms of these proxemic zones (e.g., Ferguson and Ferguson 1978). For example, close-ups and extreme close-ups may draw our attention and place particular significance on an object or person by bringing us intimately close. Medium close-ups are in the personal zone, social distances are characterized by medium and full shots, and so on. The distance from the camera to the object creates a subjective perspective that mimics social and emotional distances. Similar effects can be created with sound recording. A microphone in the same apparent distances, or proxemic zones, as those described above can create a similar effect: closely miked sounds feel very intimate, and as the sense of space in the file (more environmental sound) increases, the perceptual distance to the sound changes, and with it our sense of where the sound is physically located. The microphone is only one aspect of this perception, though, and it can be altered with effects: for instance, adding reverberation (see chapter 4) to a closely miked subject can reduce the sense of intimacy.

Exercise 3.7   Proxemics in Music

Find some examples of music you listen with closely miked vocals or guitar. What impact does that have on you as a listener? How does it compare with other songs where the distance from the sound is farther?

Exercise 3.8   Recording Distance

Record a sound at several distances, adjusting the gain but increasing the distance from the microphone to sound source, starting in the intimate zone and then

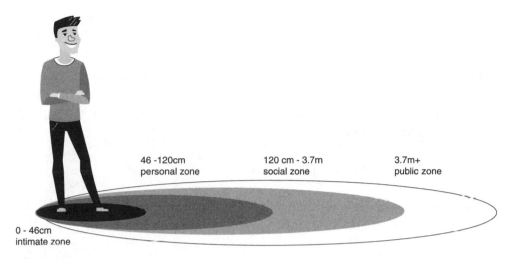

46 -120cm
personal zone

120 cm - 3.7m
social zone

3.7m+
public zone

0 - 46cm
intimate zone

Figure 3.14
Proxemic zones (after Hall 1963).

moving farther out. What are the changes you notice in the sound acoustically and perceptually?

Exercise 3.9   ASMR

You're probably familiar with the ASMR YouTube trend, and if you're not familiar, these are videos where recordings of quiet sound and whispered voice are very closely miked so as to induce an *autonomous sensory meridian response*, commonly known as tingles or chills. Intimate miking is just that—very intimate—and so it invokes a very intimate response from our bodies. Record your own ASMR session with closely miked sound effects. What type of mic did you choose, and why?

3.6   Creative Recording

The best way to get comfortable with new equipment is just to try it out in as many different situations as you can think of.

Exercise 3.10   Recording Your Daily Sound Listening Practice

Write your daily sound log. Record the sound at the same time. Then, turn up your recording and play it back: what sounds did you miss?

Exercise 3.11   Found Sound

Go to the hardware store. Find three things that make an interesting sound and buy them and record them. How many sounds can you get out of each object? Upload the sounds to freesound.org and get some feedback from users on your recording. Then repeat the exercise with garage sales (yard sales/car boot sales), a secondhand store, and a grocery store.

Exercise 3.12   Same Sound, Different Place

Record the same sound in different locations around your house and outside. What do you notice about how the sound changes?

Exercise 3.13   Moving Sound

Find a sound that is moving, such as a small creek or stream. Stand as close to the running water as possible and record the sound. Move four steps back and repeat. Kneel down and repeat again. Lie prone and repeat again, with the mic as close as possible to the sound. Play back and note your observations (adapted from Dorritie 2003).

Exercise 3.14   Repeat, Repeat, Repeat

Choose a common sound (e.g., a door closing). Record as many examples of that sound as you can in a one-hour timeframe. How many did you find? What is the greatest variation between two of those sounds?

Exercise 3.15   Creatively Making Sound (Partner Exercise)

We often use a variety of really creative approaches to making sounds. While the added visuals of movies and games helps the interpretation (which we'll look at later), the sound itself can often evoke interesting results. Try to create a sound you can't record outright (e.g., a leg breaking, a bullet flying past, a spaceship, etc.) by recording another sound in its place, and then have a partner guess what you are intending to use the sound for.

3.7   Prototyping Sounds

We often don't have the materials at hand to describe a sound we want to use, so we might use our voice to mimic the sound we're going for to share with someone or use as a temporary placeholder. This is sometimes called vocal sketching (see, e.g., Ekman and Rinott 2010). Learning how to prototype sounds verbally is a useful skill. Michael Winslow, sometimes called the "man of 10,000 sound effects," is a human beatboxer who can make all kinds of sounds with his mouth—he is best known for his roles in some 1980s movies like *Police Academy* and *Spaceballs*. Sometimes, mouth sounds are used in the prototyping phase of design. Some games have even been made with all of their sound effects created by mouth, like *Hidden Folks*.

Exercise 3.16   Making a Sound with the Mouth

Imagine someone is shoveling different materials: snow, sand, gravel, coal. How does the sound change? Make the sounds with your mouth (adapted from Schafer 1992).

Exercise 3.17   Vocal Sketching

Try making a short composition using only mouth sounds.

Exercise 3.18   Vocal Sketching Part II (Partner Exercise)

Once you've recorded a sound composition, give it to a partner and have them find recorded sound effects not made with the mouth that match what you prototyped. Alternatively, find that sound yourself and see how close you came to it, then try to get closer.

Exercise 3.19   Record a Sound Poem

Sound poems poems that are meant to be read aloud, often containing nonsense words. Some examples are "Karawane" by the Dada artist Hugo Ball, or "Zang Tumb Tuuum" by the Italian futurist F. T. Marinetti. Find a sound poem and record it with your own vocalizations.

Reading and Listening Guide

Cathy Van Eck, *Between Air and Electricity: Microphones and Loudspeakers as Musical Instruments* **(2017)**

Van Eck takes us on a history of the use of microphones and speakers in music and as music, including some early experiments with surround sound, feedback, and other techniques that are commonplace today. A must-read for anyone interested in sound art, as well as sound in performance art and experimental music.

Steve McCaffery and bpNichol, eds., *Sound Poetry: A Catalogue* **(1978)**

You can find this book floating around on the internet. It's a catalog from the Eleventh International Sound Poetry Festival held in Toronto in 1978. The introduction provides a useful overview of sound poetry in Dada and Futurist works and describes different genres of sound poetry, before the book presents some sound poems and writing about sound poetry.

There is a playlist of sound poems at studyingsound.org. You might also check YouTube for recordings: try Kurt Schwitters, Raoul Hausmann, and Jaap Blonk.

Karen Collins and Ruth Dockwray, "Sonic Proxemics and the Art of Persuasion: An Analytical Framework" (2015)

In this essay, Ruth Dockwray and I explore the role of sonic proxemics in public service announcements, and present a framework for the analysis of understanding the proxemics of sound as a rhetorical device in audiovisual media.

Gino Sigismondi, Tim Vear, and Rick Waller, *Shure Microphone Techniques for Recording* **(2014)**

This is an overview of different microphone types and techniques by manufacturer Shure. They're obviously inclined to speak mostly about their own microphones, but there is a lot of useful information about the types of microphones and techniques for recording.

4   Sounds in Space

We've talked a lot now about sounds, but have barely touched on the fact that sounds occur in space—they occur in the air, in the water, and resonate in our bodies. In fact, without this secondary element of sound (i.e., where sound occurs), sounds wouldn't occur at all! Sound cannot travel in a vacuum; or in the words of Ridley Scott's *Alien* movie, "In space, no one can hear you scream." Since sound waves travel through vibrating molecules in the air/water/solid, if there are no molecules to vibrate, the sound can't travel.[1]

In this chapter, we are going to talk about the basics of how sounds *propagate*—how they move in their environment—and what impact that has on our perception of sound. We're also going to start looking at the use of some digital effects that mimic the real-world effects of sounds in space. In chapter 7 we'll explore more advanced sound propagation along with the 3D positioning of audio in surround sound and virtual worlds.

4.1   The Doppler Effect

Stand in one spot on the side of a long, straight road and wait for a vehicle to drive by. Look down the street and listen to the sound of the vehicle as it approaches, and then as it passes and drives away. What happened to the sound? As kids we pick up on this effect intuitively and with dinky cars or toy motorcycles we make the sound with our mouths—"nnnnnneeeeeooowwww"—the frequency changes as the vehicle

---

1. In fact, this isn't exactly true. It's just that with so few molecules in space, the sound waves need to be very, very low frequency to travel from one molecule of space dust to the next. A black hole, for instance, emits an oscillation every 10 million years or so, and so creates a frequency about a million billion times deeper than we can hear! See Shiga 2010.

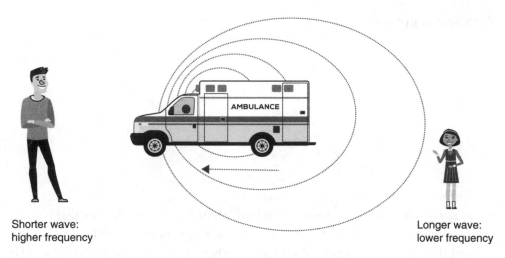

Shorter wave:
higher frequency

Longer wave:
lower frequency

Figure 4.1
Doppler effect created by a moving sound object.

approaches and drives by. This change in frequency is called the *Doppler effect* (or *Doppler shift*). You can hear it more obviously when the car is making a tonal sound, like a siren or horn.

We learned about sound waves and they how propagate in a spherical direction in chapter 1. With the Doppler effect, when the source of the sound wave is moving (as when a car is driving), the direction that it moves toward will cause a "bunching" effect of waves. For a stationary listener, the result is that as it moves toward us the sound waves are heard at a higher frequency. As the vehicle moves away from us, the wavelength becomes lengthened, resulting in the perception of a lower frequency.

Exercise 4.1   Doppler Effect

Go out and record the Doppler effect occurring on a road. It will work best with a car that has a siren on, but will work with regular vehicles too.

Exercise 4.2   Make a Doppler Effect

You can make a Doppler effect pretty easily: just get something that is making a constant sound, tie it to a string, and swing it around your head. For instance, you can purchase a cheap piezo buzzer/speaker (~$5) and attach it to a 9 volt battery.

Figure 4.2
Materials with which to make your own Doppler effect.

Stuff the noisy thing into a ball on a string. You may need to boost the signal so it's loud enough, or purchase an extraloud piezo buzzer. I found a tin Pokémon ball that unscrews into halves, but I've also used a ball designed to hold dog treats.

4.2   Reverberation

Reverberation is a massive subject, so we will cover just the basics here. When sounds move in any environment, eventually they're going to hit some form of surface, like a wall or cliff. Much like the balls on a billiards table, sound waves reflect off surfaces. The amount of reflection depends on the type of surface and the size of the soundwave, so the result may be that some frequencies are *absorbed* while others are *diffused*, and others are *reflected*.

4.2.1 Reflections

Let's first focus on sounds reflected on a hard, smooth, flat surface. When a room has a lot of highly reflective surfaces we call that a "live" or "wet" space. You probably remember in high school math class learning about reflection, where the *angle of incidence is equal to the angle of reflection*. This reflection is what happens with most sound waves: the angle that the sound approaches the smooth surface is the same angle at which it will reflect. This type of reflection, known as *specular reflection*, occurs when the wavelengths are smaller than the reflecting surface (figure 4.3).

The sound waves don't stop after hitting just one wall in a room, however: they continue, and will reflect again when they hit the next wall, if it's a very live or wet space. When sounds reflect off multiple surfaces, as in a room, we get *reverberation*, or *reverb*. Reverb consists of the direct sound, early reflections, and the late reflections, which blur or *smear* together. Each time the sounds travels, some frequencies are being absorbed by the air and by the surface they meet, and others are continuing to reflect, reducing in amplitude.

Exercise 4.3 Reflections of Sound

Use a tiny but powerful flashlight or laser light on a very hard surface (like the white board in a classroom or a mirror in your home)—turn the lights out and watch how the reflection bounces off the surface. Now direct a sound in the same place as the flashlight, and move a directional (e.g., shotgun) microphone around until you find the sweet spot—is it the same as the light? You can also use cardboard tubes to listen for and find the angle of reflection.

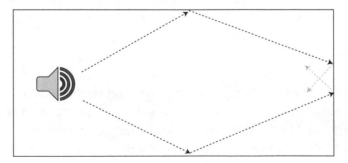

Figure 4.3
Specular reflection of a sound wave on a smooth surface.

4.2.2   Parabolas: Gathering Reflections

When the angle of the reflective surface is not flat, the sound reflects at a different angle. If the surface is smooth and convex, the sound wave will bounce outward, having an antifocusing effect (figure 4.4).

If the surface of the reflecting surface is smooth and *concave*, then we can focus the sound much in the same way waves are focused on a satellite dish. This parabolic effect is why amphitheaters are built as semicircles that reflect back to the audience, and it's even possible that some ancient spaces like Stonehenge were built in a ring for the same reason (see Cox 2014, 83–84). You may remember high school math where parabolas reflect toward a single location, the *focus point*. This is the principle of parabolic microphones: a parabolic dish focuses the sound waves onto a microphone (figure 4.5).

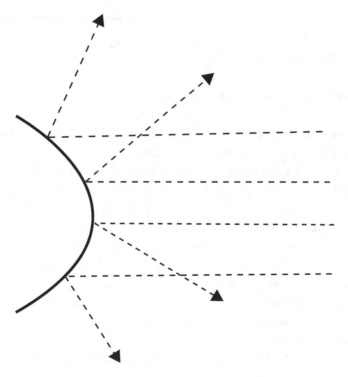

Figure 4.4
A convex surface will scatter sound waves.

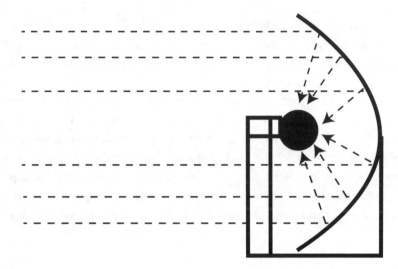

Figure 4.5
A parabolic microphone picks up the reflections from a concave surface.

Exercise 4.4   Make a Parabolic Microphone

You can buy parabolic microphone dishes that are used in field recording (a professional one will cost over $1,000), but you can also make your own. It won't be as
good, but it will be good enough to demonstrate the effect. For this exercise, we'll
need some tools. First, we need a large bowl as close to a parabola shape as possible
(if you can find a discarded TV satellite dish, that works great!). It needs to be a
highly reflective, perfectly round dish, like a plastic salad bowl. In an ideal world
the bottom of the bowl would also be round, but most salad bowls you'll find will
have a flat bottom. We know that the size of sound waves increase as we lower
the frequency, so larger dishes will reflect more sound: You may see small cheap
tiny parabolic microphones on a handheld gun-like device on eBay, for instance,
but these will only reflect high-frequency sounds. The larger the dish, the more
frequencies we will amplify.

   Next we need to find and mark the focal point. Drill a ¼" hole in the center of
the bottom of the dish and stick a dowel or chopstick into it, straight into the bowl,
with the length of the stick longer than the height of the bowl. This will be our
measuring stick and will hold our microphone. If you've got a laser pen, you can
shine it onto the bowl and it should reflect at the focal point. Check a few different

angles to be sure. Mark the focal point with a pen. If you don't have a laser pen, you can use your ears and the microphone. Set up a directional sound and point the bowl toward the sound, and move the microphone until the amplitude is at its loudest.

Tape a small omnidirectional microphone, like a lavalier microphone to the focal point, facing the bottom of the bowl.

4.2.3   Flutter and Standing Waves

Most rooms have parallel walls, and that means that reflections of particular sound waves will create some interesting phenomena. *Standing waves* are waves that reach a barrier such as a wall, and bounce back at the same wavelength, building up by bouncing back and forth. The result is a sometimes visible disturbance of the air. Some people have theorized that standing waves could have been created by humans singing in caves where ancient people have painted images, and perhaps the reason for the caves holding special purpose for early humans was due to their acoustic properties (see Coimbra 2016). As the wave bounces back and reinforces itself, a process of constructive and destructive patterns occur, creating what are called *nodes* and *antinodes*. Nodes are where the wave stays in a fixed position because of destructive interference. Antinodes are the places on the wave where it vibrates at the maximum amplitude. Standing waves appear to be "standing still" in the space as they bounce back and forth. If we use our analogy of the billiards table with a ball bouncing off the walls, there are places where the ball will always repeat: if that ball were traveling fast enough, it would appear to be standing still in those places (figure 4.7).

Chladni (pronounced "klad-nee") plates demonstrate the standing wave patterns that are created by the constructive and destructive interference of waves. Salt or sand is placed on a plate and vibrated at a set frequency. The sand gathers at points along the plate that are not vibrating, creating patterns. These are the same types of constructive and destructive patterns that make a great violin (figure 4.8).

If the time it takes for the sound wave to bounces between parallel walls is more than 50 ms, what sometimes results is a pattern of reflections known as a *flutter echo*, sometimes called "chatter" or "zing," although many people describe it as a "boing" sound. Flutter echoes can make the sound appear to "flutter," and sound quite hollow or "ringing." You can often hear a flutter echo if you go into a smooth, highly reverberant large space, such as an empty gymnasium, and clap your hands.

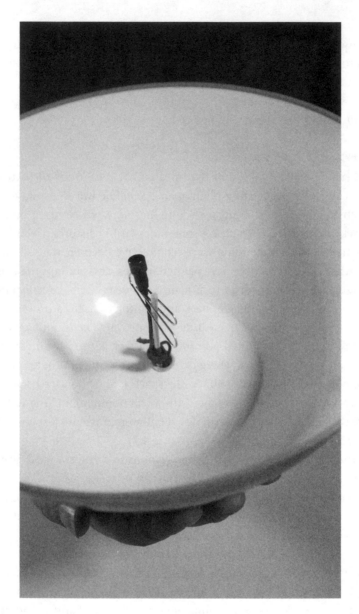

Figure 4.6
A homemade parabolic mic with salad bowl.

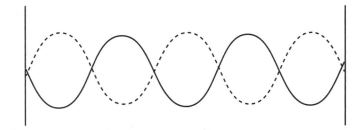

Figure 4.7
A standing wave created by the reflection of a sound wave off of perpendicular surfaces.

Figure 4.8
Chladni plate patterns.

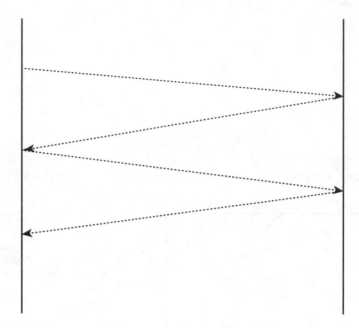

Figure 4.9
Flutter echo created by parallel walls.

Exercise 4.5   Standing Waves and Flutter Echo

Visit some different indoor spaces around your campus, house, mall, etc. Clap your hands and listen for the reflection. Try to generate some frequencies between parallel walls and see if you can create a standing wave pattern. You will hear more reflection if you can go at a time when there aren't many people around. Did you find anywhere that has a flutter echo? Did you find anywhere with a standing wave?

4.2.4   Resonance

Resonance refers to the constructive and destructive interference pattern of sounds in an object: All objects have frequencies at which they vibrate naturally, which are called *resonance frequencies*. The shape and size of a resonant cavity means that some frequencies interfere constructively and create standing waves, while others interfere destructively and are canceled out.

If you've ever seen a cartoon where an opera singer breaks glass, this actually works, and is to the result of resonance. It needs to be crystal, not glass, to work effectively. With a finger dipped in water, rub your finger around the top of the glass, and you'll make the glass sing (hum). It's possible to use different-sized glasses to create a musical instrument—these are known as glass harps, glass harmonicas, or armonicas. Benjamin Franklin invented his version in 1761, which included thirty-seven different crystal bowls. There were rumors that the playing of the instrument would cause madness and depression in both player and audience, so the instrument fell out of favor, although some more recent artists like Björk and David Gilmour (of Pink Floyd) have used the glass harmonica.

Exercise 4.6   Singing Glass

Record the sound of a crystal glass humming, and look at a spectrogram to determine the frequency. It is possible to break the glass by playing a sound that is the same frequency and the right amplitude to exceed the strength of the crystal. (You can usually pick up used crystal glasses at charity shops for a few dollars.)

4.3   Absorption and Diffusion

If you've ever heard that some people put egg cartons on the surface of their home recording studio, it's down to the belief that the egg cartons work as a *diffuser*. Instead of the sound waves reflecting back evenly, the theory is that the waves will tend to scatter, reducing the amount of reflection (reverberation) in the room. In fact, egg cartons don't actually work that well. The *absorption* ability—called the *absorption coefficient*—of cardboard and Styrofoam aren't very effective. Absorption occurs when the surface material absorbs sound energy rather than reflecting it. This is why sound studios are commonly treated with some type of absorption material, such as foam. Absorption will reduce the reverberation, or "liveness," of a room.

Professional sound studios designed for recording music often want to retain some degree of liveness in a room, so they will use *diffusers*. Diffusers scatter the reflections rather than absorbing them. A diffuser needs to be at least six inches deep to be effective, and needs to be about six feet away from microphones or sound sources, since they introduce artifacts into the sound. Most professional studios will use a two-dimensional diffuser, which looks a bit like bricks of many different lengths. So, in addition to not absorbing sound very well, egg cartons are also not effective diffusers, since the egg crate cups are all at the same short depth, and will only diffuse a narrow range of frequencies and not others.

If you look up at the ceiling of a classroom, doctor's office, or other public space you'll often notice tiles with lots of different-sized holes in them. These are acoustic tiles, designed to absorb and diffuse sounds, reducing reverberation to increase clarity of speech. Low-frequency sounds are absorbed, the deep cuts are designed to absorb and diffuse mid-frequency sounds, and high-frequency sounds pass through the holes to be dispersed in the space inside the ceiling.

It's not always possible to install absorption or diffusion tiles where we need to record, but we can do some things to improve our recording. If we are recording in a space with a lot of reflective surfaces, we can hang some cloth to reduce the reflections to get a clearer sound. If you're recording something small or your voice, you can make a small vocal-booth-style box with a homemade frame using PVC piping from a hardware store and some heavy blankets, or specially designed acoustic blankets.

Sometimes we want some reverberation: we tend to associate a little reverberation with warmer sound. You may notice you sound better when you sing in the shower, because you're hearing the reverberation in the shower stall come back at you with your voice, thickening out the voice. For a time in the late 1950s and 1960s, it was very trendy in rock recording to have a big reverberant sound, and singers like Duane Eddy, Elvis Presley, and Lee Hazlewood were recorded inside large oil drums or grain elevators. Stax Records even recorded their stars in the bathroom of the studio (Zwisler 2017)!

Exercise 4.7   Same Sound, Different Place

Record the same sound with the same microphone at the same distance and angle in different places inside and outside. Hang some cloth around the sides of your mic and notice how it changes the sound of the recording. Try to find a highly reflective space like a four-sided glassed-in shower, or a concrete or plastic playground tube.

Exercise 4.8   Room Tone

Room tone is typically gathered on a film or television set for later use, and involves two minutes without anyone making noise, just recording the ambient space (inside or outside). Go record some room tones!

Exercise 4.9   Stitching Reverbs

Play your favorite song in three different rooms on the same device, and record the playing of the song in the space. Put ten seconds of each version back to back a few times, alternating with hard cuts between them, compare the sound of each.

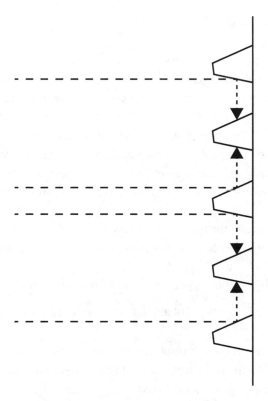

Figure 4.10
Sound diffusion.

4.4 Digital Reverberation

Rather than recording in our bathrooms, normally we want to record in a "dry" or fairly "dead" space and add reverberation digitally afterward, so that we can adjust the amount of reverb to what we want. Reverb can simulate a space, but it can also add a little texture, creating a soft, ethereal feeling, and it can be used to provide contrast (e.g., bands might put reverb on some of their instruments but not others). Reverb can be used on sounds to create a sense of flashback, in-the-head, or "voice of God" effect. We can also use just the reverberation of sound, known as the "tail" as a sound effect in itself, creating otherworldly sounds, similar to taking the attack portion off the sound as we did earlier.

Often there are presets for reverb that allow you to choose the type of space effect you are going for. But understanding the basic settings will help you to adjust your own reverb settings.

Usually the settings will involve some or all of the following:

Reverb time refers to the "liveness" of a space—how much time it takes for the sound to reflect and fall to silence.

Bass ratio is related to this reverb time, in that the spectral content of the continuing signal changes with the time, with more of the higher frequencies rolling off sooner because of the length of the sound waves.

Pre-delay is the difference between when the direct sound occurs and when the first reflection occurs (i.e., the time between the direct sound and the *impulse response*). Sometimes this is calculated as the time between the direct sound and the late reflections.

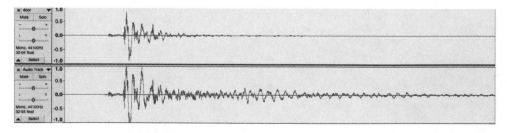

Figure 4.11
Comparison of dry (top) and sound with light reverb on it. Note the longer tail on the second sound.

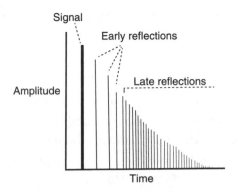

Figure 4.12
Reverberation: original signal and reflections.

In Audacity, the additional parameters are as follows:

Room size refers to the size of the simulated room, from 0 being the smallest size to 100% indicating a very large room.

Reverberance is the length of the reverb tail—how long it takes to end (reverb time above).

Damping stops the reverb from building up too much; also, high frequencies will decay faster (i.e., the high frequencies are damped out).

Tone low reduces the low-frequency elements of reverb, making it less boomy.

Tone high reduces the high-frequency elements of the reverb.

Wet gain increases the volume of the reverb (wet) component.

Dry gain increases the original sound (i.e., the dry sound).

Stereo width, only in stereo sounds, sets the perceived width of the stereo field by adjusting the amount of reverb in each channel separately.

Wet only means that the original sound will be removed and only the reverb tail will be heard.

Reversing reverbs can create a "backward" sounding effect, particularly if a reverse reverb is layered below a forward-played effect, creating a *preverb* effect. This can be used on vocals to create a psychedelic, or ghostly effect. To create the effect:

(1) Duplicate a sound file.

(2) Reverse the second file (select the track, then Effects > Reverse).

(3) Apply a reverb with a long tail (like a large room), using `Effects > Reverb > Manage > Factory presets > Large room.`

4) Reverse track 2 again (`Effects > Reverse`).

5) Mix and render.

You may also come across the term *convolution reverb*. Convolution reverb applies all aspects of the reverb patterns of a space onto a sound. A real space is typically measured using impulse responses—calculating the distances and times of frequencies in a reverb pattern to record the "signature" of a space. It is possible now, in other words, to appear as if your file was recorded in Carnegie Hall by applying convolution reverb to your dry sound file. You can download impulse responses from libraries just as you might sound effects. Some are free and some are paid.

Exercise 4.10   Reverb in Audacity

Record some sounds and try out some different reverb effects on them. If you select the "Manage" menu, you will be given the option of a variety of presets. Try the different settings out and preview the setting to hear the difference. Try some preverb effects as well.

Exercise 4.11   Thinking about Reverb

What associations or memories do you have with reverberation? How much reverberation do you associate with warmth, versus the "voice in my head" effect? To create the voice of a ghost? To simulate a sound in a cave? Try using reverb on different sound samples and at different settings to create different effects.

4.5   Echo and Delay

In addition to reverberation, some reflections are delayed so long that they are called an *echo*. In technical terms, an echo is a distinct signal that comes back with a delay of more than 1/10 of a second (0.1 seconds). Sound travels in dry air at 20° Celsius at a rate of 343 meters per second. The distance a sound will travel in 0.1 seconds, then, is 34 meters. So, if we stand at least 17 meters from a barrier and make a sound (i.e., the distance from us to the barrier and back is 34 meters), we should hear an echo. If we stood closer to the barrier, we would instead hear the reflection as reverberation, not an echo (i.e., not a distinct sound reflected).

Delay effects take a sound signal and delay it before playing it back: a delay of more than 0.1 second effectively creates an echo. As such plugins are sometimes separated out into delay and echo, but usually we don't need separate delay and echo plugins. In music production, delays are often set to the time of the music so that they are intentionally on-beat or off-beat, so many delays have not a specific time but a beats per minute, or bpm, setting. The effect of a delay can be to create a sense of space, create psychedelic effects, to create a sense of "liveness" or thicken out a sound.

Most delays have some kind of *feedback gain control* which keeps sending the signal through the delay, reducing the amplitude each time. The plugin takes the signal, spits out one sample, then feeds it back into the delay again, but at a reduced amplitude. There are straightforward fixed delays and there are variable delays that modulate the amount of time (rate) the delay takes (to vary the time between signals). Variable delays may also vary the *shape* (the path taken by the delay setting in changing its parameters) and the depth of delay (how much the delay is modulated). Figure 4.13 shows how a delay plugin works: the original signal (dry) is split and the signal is sent to the delay (becoming "wet") where it is delayed by a certain time, then output. The output is often split into two and fed back into the delay. The dry signal is then mixed with the wet for the final output.

4.6   Digital Delay

If we take a sound sample and add some delay, adjusting the delay time so we that have a full sample returning to us, we can see that the delay is processed so that each time it comes back it has a reduced amplitude (figure 4.14).

Open a sample in Audacity. Select `Effects > Echo`. There are two options in the pop-up window: the first is the time between the dry signal and the echo, and the second is how much of an amplitude decay between echoes there should be (on a scale of 0 to 1). Note that the number of echoes you get will depend on the length of your

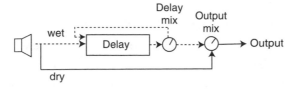

Figure 4.13
Delay diagram signal path.

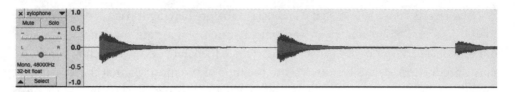

Figure 4.14
Digital delay in Audacity.

file—if you don't have enough room in the file for the echo, it will get cut off, so you need to add silence to the end of your file by clicking on the end of your sample, and selecting `Generate > Silence`.

Go back to your original (dry) file. Now open the `Effects > Delay` panel. Audacity offers several options:

Delay type: regular (each time it repeats it has the same delay time), bouncing ball (makes the delay time decrease in time as it repeats), and reverse bouncing ball (increases the delay time between repetitions).

Delay level per echo: the amount of change in amplitude in each delay (in dB). The default is negative, meaning it will get quieter. Positive values may cause the sample to clip, because this is the amplitude you are adding to the original signal.

Delay time: the time between echoes. As we know from above, reducing the time can create a reverb rather than a distinct echo effect.

Pitch change effect: pitch/tempo. If you want the delay to change the frequency, you can select it here. Changing the frequency changes the wavelength, which changes the time, so it is also a tempo effect.

LQ pitch shift: changes the frequency while maintaining the same tempo. The quality of the sample will degrade.

Pitch change per echo (semitones): a semitone is half of a full tone on a musical scale. Two semitones will change the frequency to a full tone, in other words. You can also set to fractions of a semitone.

Number of echoes: how many times the delay will sound.

Allow duration to change: if you start with a short file time that ends at the end of your sample, you won't have room in the file to accommodate the delay, so if you want to accommodate the delays you would select "yes" here to increase the file length.

Exercise 4.12 Echo

Try out some different echo settings (Effects > Echo) on a sound sample, and listen and compare. What uses besides environmental sounds could you imagine for an echo effect?

Exercise 4.13 Digital Delay

Try out the various delay settings and presets on a sound sample, and listen and compare. How does the delay effect compare with echo?

Exercise 4.14 Hearing Delay

Find three examples of music that use delay effects. Why do you think the musicians chose to use delay? What effect does it have?

4.6.1 ADT and Chorus

There are additional delay effects that you will commonly encounter, including ADT and Chorus, although you won't find them in Audacity without downloading additional plugins.

Chorus is when a signal is duplicated, usually multiple times, with a slight deviation in delay, with the amount of delay modulated by a low-frequency oscillator, creating a copy that is slightly varied in pitch, which is then mixed back with the original. The effect is usually used to create the feeling of a chorus, or multiple people singing very slightly off-pitch from each other and very slightly out of sync.

ADT, or automatic double tracking (sometimes called "artificial" double tracking), is used for filling out a voice or instrument, often with some other effect on the signal. It is effectively a chorus effect created by delaying an audio signal and then mixing that back with the original. ADT is an automated process based on a historical approach to singing multiple takes and mixing takes together (double tracking).

Exercise 4.15 Chorus

Create a chorus effect in Audacity by duplicating a single sample three times, giving it a slight delay each time—in effect, you are manually double-tracking:

(1) Duplicate a sound file. Do this twice, so you have three tracks.

(2) Use the transport icon to adjust the start time of the second and third tracks, altering the delay.

(3) Change the frequency of the delayed tracks slightly by using `Effects > Change Pitch`.

How does this chorus style effect change the ways that you hear the sound now? Where do you think you might use an effect like this?

4.7 Phasing and Flanging Effects

We learned about sound wave constructive and destructive interference in chapter 2. A *phaser*, also known as a *phase shifter*, uses these properties of sound waves. A phaser splits the sound into two signals, where one uses a filter that shifts one of those signals out of phase, resulting in constructive interference when the peaks overlap, and destructive interference when both the peaks and valleys overlap. With a phaser, the switch back and forth into and out of phase is created with a sweep across the frequencies, creating a "whooshing" sound.

A *flanger* is a type of phaser that also uses a second delayed signal to modulate the original, but there are technical differences in how the result is produced. With a phaser, the phase of some frequencies may be shifted a different amount than the phase of other frequencies, so the peaks and valleys of the output aren't necessarily harmonically related. A flanger on the other hand uses a delay evenly across all frequencies: The out-of-phase "whooshing" effect is therefore more noticeable when we use a flanger.

We can see the difference on the waveform if we generate a sinewave tone then apply the phaser effect. Although the pattern repeats (see figure 4.15), there are variations within the pattern. Compare that to the waveform of this flanger effect (see figure 4.16). This flanger is an *even repetition* of the pattern in the effect (note also the boost in amplitude in this flanger).

Figure 4.15
Phaser effect on a sine wave.

Figure 4.16
Flanging effect on a sine wave.

Exercise 4.16   Listening for Phasing

Find some examples of music that use a phasing effect. How does it change the feeling of the sound?

4.7.1   Digital Phasing

The phaser in Audacity uses an *LFO* (a low-frequency oscillator) to shift the phase. Altering the frequency will alter the rate at which the sweep of frequencies occurs. Open a sample in Audacity. It may work better if your sound file is at least a few seconds long so you can hear the effect more easily. Click on `Effects > Phaser`. There are several options:

Stages: we know phasing creates a constructive and destructive interference pattern. The more stages you have, the more peaks and valleys in the response: effectively, the stronger and more complex the phasing effect will be.

Dry/Wet: 0 is the unprocessed signal, and 255 is only the delayed signal, so the best mix of the phasing effect is half-way in between, at 128. (Sometimes this is represented as a percentage, where 50% is the most effect heard.)

LFO frequency: as discussed above, the second signal is what shifts the phase. The LFO frequency will adjust the sweep rate across the frequency range.

LFO start phase (degrees): the LFO sweep starts at 0 by default and sweeps from high to low. At 180 degrees, the filter sweeps up from a low frequency.

Depth: how high the frequency sweep is: Higher settings alter from low to higher frequencies. Lower settings will affect only the low frequencies.

Feedback (%): increasing the feedback will feed the signal back through the effect more times, creating a more obvious phasing effect. Negative settings reverse the signal and will also create an interesting effect.

Exercise 4.17   Phasing

Explore the phaser effect by adjusting the phaser parameters and presets on a sound you're familiar with. What do you think you would use different degrees of this effect for?

Exercise 4.18   Flanging

There is no flanger effect by default in Audacity, but you can install a plugin flanger effect and explore flanging. How does it change the sound of different samples?

4.8   Time-Stretching

Stretching a sample out over a longer period of time is called *time-stretching*. Time-stretching may or may not change the frequency of the sample, depending on the plugin. Technically, changing the length of time a sample plays with raise the frequency when shortening the time, or lower frequency when lengthening the time, unless the effect allows for simultaneous frequency adjustment.

In Audacity there are two ways to stretch out the time. The first is to change the tempo (`Effects > Change Tempo`). Here you will see the option to change the frequency (pitch) at the same time. This particular plugin is set up for music, so it gives the option in beats per minute, but it also has the option to change length in seconds. This will create a stuttering. The second option to alter the time of the file is to change speed (`Effects > Change Speed`). This will give you a smoother stretch (figure 4.17).

Exercise 4.19   Droning On

Use the speed and tempo tools (`Effects > Change Speed | Effects > Change Tempo`) to make a drone from an everyday sound. Try to stretch it out to different lengths, and try out different sounds.

4.9   Worldizing

Sound designer Walter Murch uses the term *worldizing*, which refers to creating a place or space in which sound exists, a world in which it inhabits, taking into account the acoustic atmosphere and the role of space in the perception of sound. It seems an obvious part of designing sound, but it hasn't been all that long since these techniques

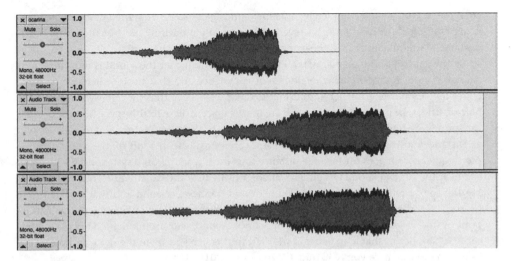

Figure 4. 17
Sample dry (top), speed (middle), and tempo (bottom).

began being used in film sound, for instance, or in game sound. I interviewed Nick Wiswell, who designs the sound for racing games, and he explains the idea in detail:

One of the big things for me, when simulating the sound of a race, is simulating the environment that the car is driving through. If you think about in the real world, you've got a car, you've got your barriers, you've got your things behind the barriers, you've got the general ambience of the world. Sometimes you've got tunnels. And if you've got a nice-sounding car, the first thing you do when you put it in a tunnel is you wind the windows down, you slow down as much as you can, and you floor it, just so you can hear what it sounds like. We want to simulate that.

But you also get those little sensations if you're driving along and then suddenly a concrete barrier is alongside you, a foot from the side of the car, you'll notice the sound completely changes as you hear the tires and the engine slap back off that wall. One thing I think racing games can focus too much on is the engine sound. You play a racing game and you hear it and it's just engine. That's all it is, is engine. And it's like, well, surely the interaction of the way the car travels through the world should be just as important. Make it feel like a space. On the graphic side they start talking about HDR lighting and image-based lighting and the lighting reacting to the world around where you are. I want to reproduce that from an audio perspective, so we're building a system in the game that allows us to model various early reflections and more distant reflections actually real-time. It's not baked into a reverb effect. You'll actually hear, as the car approaches a barrier, you'll hear the sound start to reflect from the barrier. Behind the barrier, you could have buildings and then no buildings and you should hear that sound change in contrast as it does so. You'll enter a big grandstand area and you should hear the sound sort of echoing around in there. Drive through a tunnel and you're completely enclosed, so you want a completely different sound again, so we've started designing systems using multiple delays and multiple reverbs to start really modelling not just the sound of the car but the sound of the car in the world around it and how it changes.

So the car itself isn't changing, but everything around it that the sound is reflecting off is changing, and that creates a whole different experience and that's somewhere where we're going to keep pushing further and further because we believe that's where we can make the biggest strides, in making it feel like you are in this space. You're in this space not just because the sounds are right, but how the sounds are reacting to the environment and how far away it is. Distance models on a lot of things in games aren't realistic. It's sort of, the sound is gone, pretty quickly as it goes off into the distance, but we're working on a system now where you can hear the car two kilometers away. You don't really hear the car anymore, but you're hearing the sound reflecting off the environment and bouncing around and that's what you're hearing. You're not really hearing the direct sound of the car. You're hearing all the interactions of the car with its world and that all bounce around, and I think that's a big push. If we can start simulating that a lot better, it will start feeling like you are in the space and the sound is reacting to the space you're in. (quoted in Collins 2016, 312)

Although both Walter Murch and Nick Wiswell are speaking specifically about sound design to support visual media, the concept of worldizing is equally—if not more—applicable to sound design for audio stories. We'll come back to this point in chapter 9, but identifying the location of a scene through sound is a key role that sound design plays in radio dramas and fiction podcasts. Worldizing can involve applying impulse responses of real spaces to make the sounds feel more realistic in a particular environment, or it can be recording sounds in particular spaces to capture the change in sound, or using reverb effects creatively. Whatever technique you use on a sound should be used on all sounds in that particular scene, to create a very realistic sense of the space, unless you're trying to have certain sounds stand out by reducing or eliminating the reverberations.

Exercise 4.20   Worldizing

"Worldize" a group of sounds into three different locations: how does it change the way you hear the sounds? Try to use each of the techniques: record the sounds in one space, use the same reverb settings, or download an impulse response for a space and apply that.

4.10   Setting Up a Recording Space

If you don't have access to a professional recording space, now that you know a little bit about acoustics it's often possible to set up your own at home fairly cheaply, which will suffice for most beginner purposes. Many videos and tutorials can be found online, and numerous books are available about setting up a home studio on a budget. Here are a few tips:

1. Turn off any appliances and HVAC while recording. You don't want your heating or refrigerator suddenly kicking in. You may be so used to hearing those sounds that you don't notice them.

2. Reduce reflections where possible. You can find instructions online on how to build an affordable vocal booth with just a few parts from the hardware store and some blankets. Vocal booths also suffice for most indoor recording. You can purchase acoustic foam, acoustic panels, and sound blankets to temporarily hang over windows while recording if you can't build a dedicated home studio. Don't forget the floors—a rug or carpet is going to be an improvement over hardwood or tile that can reflect sounds and cause phase cancelation issues.

3. If you can afford it, bass traps in the corners of the room can help to eliminate some issues.

Reading and Listening Guide

Peter Doyle, "From 'My Blue Heaven' to 'Race with the Devil': Echo, Reverb and (Dis) ordered Space in Early Popular Music Recording" (2004)
In this article, Doyle talks about the introduction of room ambience to popular music recordings, comparing the techniques in music with film sound design of the era and tracing the techniques through the various changes in recording studios.

Fernando Augusto Coimbra, "Neolithic Art, Archaeoacoustics and Neuroscience" (2016)
This may be an academic paper with a lot of big words in the title, but it's an accessible and fascinating read about acoustics in prehistoric times that brings together a lot of what we've learned in this chapter.

Charles Maynes, "Worldizing: Take Studio Recordings into the Field to Make Them Sound Organic" (2004)
Maynes talks about Murch's approach to his remake of the sound for Orson Welles's *Touch of Evil*, along with other recent approaches including Richard King's sound work on *Gattaca* and others. A short, accessible read.

Miguel Isaza, "Walter Murch Special: The Concept of Worldizing" (2009)
A video of Walter Murch talking about Worldizing and how he came about using it:
http://designingsound.org/2009/10/07/walter-murch-special-the-concept-of-worldizing/.

NPR, "Conquering Reverb: Behind Recorded Music's Oldest Sound Effect" (July 6, 2012)
A short (five-minute) NPR program about reverb in popular music.

5   Sound Effects

We've already looked at some space- and time-related sound effects, or *digital signal processing* (DSP) effects. But there are lots of other digital effects we can play around with that have different impacts on our sound. A simple analogy is with photos and Instagram filters: you have an image, but then you use filters, or effects, to modify the image to enhance or alter that image. We can create different feelings or enhance or alter a sound by adjusting the sound with effects. Effects can add a lot to our subjective experience of sound, and are a staple of sound design. In this chapter we will focus on frequency effects, such as modulation, equalization, and filters. There are other effects that we're not going to cover in this book, and you'll find yourself gathering software plugins over time that are designed for specific purposes, but we'll cover some of the most common effects. I encourage you to download and explore other effects available. Now that you have a basic understanding of sound waves and digital sound, the documentation that each effect has on the Audacity website should start to make some sense.

5.1   Tremolo and Vibrato

Tremolo modulates the amplitude of a sample back and forth, giving it a "trembling" effect (*tremolo* is Italian for "trembling"). You can modulate the speed at which the tremolo works, and also the *depth*, or how much amplitude is adjusted. Tremolo can give a feeling of "liveness," pulsation, or movement (figure 5.1).

5.1.1   Digital Tremolo
You may need to download the tremolo plugin to use tremolo. Generate a sine wave as we have done many times now (`Generate > Tone`). Apply the tremolo effect by selecting `Effects > Tremolo` (just select the default for now and hit OK). Notice the

Figure 5.1
Sine wave tremolo effect.

amplitude of the wave has changed to a wavy pattern, although the frequency is the same (figure 5.1). Open the Plot Spectrum (`Analyze > Plot Spectrum`) to check that there is no change in frequency.

Audacity has several settings for tremolo:

Waveform type alters the shape of the amplitude pattern: triangle creates a triangle wave pattern and square creates a square wave pattern.

Starting phase sets where to start in the waveform cycle, so zero is the start of the cycle (as the waveform rises toward its peak).

Wet level sets the depth of the amplitude sweep from zero to maximum of the original amplitude volume.

Frequency controls the speed of the tremolo's oscillation (how "trembly" it is), not the frequency of the file.

Exercise 5.1   Tremolo

Try some different tremolo effects on different types of sounds. How does it change the feel of the sound? Can you find examples of tremolo effects in some music you listen to? Why do you think the artist chose to use tremolo in those songs? See if you can find an example of its use on vocals!

5.1.2   Vibrato

Tremolo is often confused with vibrato. *Vibrato* modulates a *frequency* back and forth (from the Italian *vibrare*, "to vibrate"). Vibrato gives a pulsating change of pitch, and can add depth and interest as well as added emphasis or expression. The "whammy bar" on an electric guitar is a vibrato effect, although sometimes people call it (wrongly) a tremolo arm.

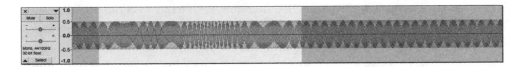

Figure 5.2
Vibrato on a sine wave.

5.1.3   Digital Vibrato

You'll need to download an extra plugin to get the vibrato effect in Audacity. Apply the effect using the default settings (`Effects > Vibrato > OK`). This plugin will currently only allow us to apply vibrato to a small sample (~2.3 seconds), so you will have to select a small part of the waveform. Unless we've zoomed in, it won't look like anything has changed, but listen to the change. If we zoom in, we can see in the waveform how the vibrato effect has changed the waveform: the amplitude remains the same but the wave*length* of the wave (i.e., the frequency) has shifted (figure 5.2).

Exercise 5.2   Vibrato

Try out some vibrato on different samples. Where do you think these might be useful for sound effects, rather than music? Find some examples of vibrato used in music. Why do you think the artist chose to use vibrato in those songs?

5.2   Pitch Shifting and Auto-Tune

A pitch shifter does what its name implies—it shifts the pitch (the musical term for frequency) up or down. Pitch shifting in music is most commonly used in association with other effects, like ADT or chorus, or to correct a singer who is slightly off-pitch, but it can also be used for all kinds of sound design purposes, like creature vocalizations. We know that changing frequency will change the length of time, since the oscillations become tighter or longer. A sound of 60 Hz (remember, that's cycles per second) altered to be 30 Hz (30 cycles per second) should be twice as long, because the cycles are being stretched. Audacity's pitch change plugin will attempt to adjust the time length to match the original while still changing frequency. This can create sometimes interesting, and sometimes unwanted, auditory artifacts (figure 5.3).

An obvious pitch correction with only part of the signal corrected results in an effect that sounds like an Auto-Tune (since this is what an Auto-Tune does!), and has more recently become a stylized popular vocal effect to deliberately distort vocals.

Although it's used generically to describe that obvious pitch correction, initially made famous by Cher's "Believe" (1998), Auto-Tune is a brand name, a processor developed by Antares Audio. In the case of the famous Auto-Tune effect now common on songs, the pitch correction speed is set too fast, creating that little distortion in the adjustment. Auto-Tune takes a few milliseconds to kick in on each note, giving the voice the slightly swirly sound.

Exercise 5.3   Auto-Tune

Find three examples of songs that use Auto-Tune: it shouldn't be hard with pop songs today! Why do you think the artists are using the effect—what impact does it have on the listener?

Exercise 5.4   Pitch shifting

Take three sound samples and pitch them up or down in small or large increments (`Effects > Change Pitch`). Audacity gives two options for changing frequency. The first is for musical pitch, and it will guess the pitch your sound is closest to, and ask you what pitch you want it to be. You can use this, or use the second half of the window below where you can deal directly with frequency (which is more detailed, granular correction). How might you use this effect in your own sound work?

5.3   Equalization

An *equalizer* or EQ adjusts the amplitude of specified frequencies within a track. Whereas volume knobs boost the entire sound file, an EQ will let you boost or attenuate (reduce) frequency bands within that sound file. Imagine if we recorded a track and found that the bass seemed flat. We can EQ the mix so the bass is increased. Remember our frequency response charts (section 1.3.1)? If we know there is a dip in the 500 Hz range of our output, we can EQ it back up. You've probably used EQs on your stereo system, as a simple knob that adjusts the bass and/or treble, or as faders (slider inputs) (figure 5.4).

There are different types of EQs, and they can vary greatly in precision or visualization, but commonly have some of the same properties. The goal of an EQ is to modify the frequency spectrum, by boosting or attenuating specific frequencies or groups (*bands*) of frequencies. These are adjusted on a slope, rather than a harsh notch of frequencies, so if you pick 500 Hz to boost, it will boost on a slope up to and down away from 500 Hz, not *just* 500 Hz.

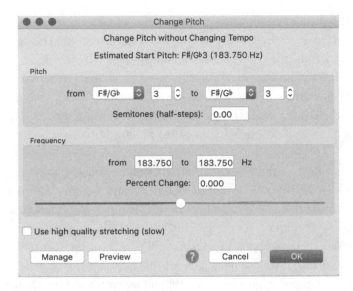

Figure 5.3
Change pitch plugin in Audacity.

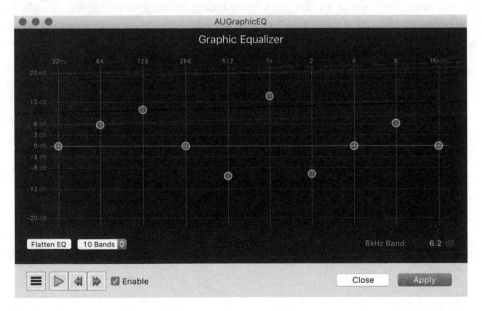

Figure 5.4
EQ faders as you'd commonly see on a home stereo system.

The slope is typically set using dBs per octave (a musical term representing a range of frequencies): the greater the dBs, the harsher or tighter the slope. The *bandwidth* is the set of frequency bands (the width of bands) being adjusted by the slope. A wider bandwidth will boost a wider range of frequencies to either side of the selected frequency.

The Q *value* is a ratio of the center frequency to bandwidth: as we increase Q, we lower the bandwidth, or sharpen the range what is being adjusted. The Q value is not shown in the standard Audacity equalizer plugin, but you may find it on other equalizers.

If we open up the first standard equalization effect in Audacity (`Effects > Equalization`), we will see something that looks like a blank Fletcher–Munson graph. The graph is flat because this is your starting point. It is not a spectrogram, so it does *not* represent where our frequencies are currently in relation to each other. There are a number of presets we can apply under the `Select Curve` dropdown. Select the `Telephone` preset and take a look at your graph now. It is showing that anything below about 130 Hz is going to be reduced to –30 dB and we won't hear it. There is a slope up to 400 Hz where we will hear everything in our original file at its original amplitude, and then at about 2,500 Hz it begins to slope down—or drop off—again. Preview the file and listen. We can adjust the curve by dragging the tiny Bezier points. Beside EQ Type on the bottom left, select Graphic instead of Draw, and we'll see some fader sliders: now we have no Bezier points but we can adjust frequencies by moving the sliders.

Exercise 5.5   EQ Presets

Try some of the different presets (select curve) built in to the Audacity equalizer. What does each preset do? Write down the effect it has on the frequencies as you both hear and see them.

Exercise 5.6   The *Best* Frequencies

Grab a long sample or one you've recorded of some length and play with the EQ until you think you've made a significant improvement to the file. What frequencies did you boost or cut and why?

5.3.1   Digital Wah-Wah

The *wah-wah*, sometimes known as the "Jimi Hendrix" effect for his heavy use of this technique, is a variable EQ, where the boosted frequency alters according to a *sweep* of the frequency band. The wah-wah is actually an instance of onomatopoeia, based on

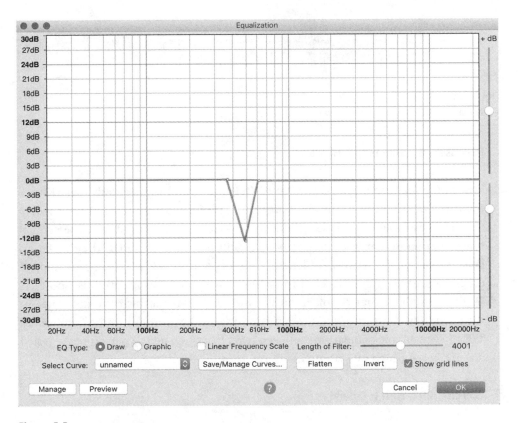

Figure 5.5
Audacity Equalizer adjusting one tone, 500 Hz, but with bandwidth of about 200 Hz.

the effect it makes. The original intent was to mimic a trumpet using a mute (a device stuffed into the end of the trumpet to adjust the amplitude and to a lesser extent the timbre/tone). A wah-wah gives a weird in-and-out wave effect as the frequencies are swept, using a low-frequency oscillator, or LFO. If we apply it to a single 500 Hz sine wave, we get a really weird-looking wave-form (figure 5.6): in this you can see the amplitude moving in and out in a wah-wah-wah-wah.

Exercise 5.7   Wah-Wah-What?

Apply the wah-wah to some sound samples and note what you discover about the sound and its effect on you. Try to find some musical examples of the wah-wah effect: usually you'll hear it in psychedelic rock or funk from the 1970s.

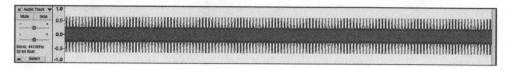

Figure 5.6
Wah-wah effect on a 500 Hz sine wave.

Exercise 5.8   Supersonic Hearing

You may notice the EQ band goes up to 40 KHz on professional equivalent. Knowing that human hearing cuts off at about 15 KHz for most adults (20 KHz if you're young), why would the bands go up to 40 KHz? This is a mysterious perceptual phenomenon in that even though people can't hear above 15 KHz, they appear to hear the difference between a file where the upper frequencies have been chopped off and a file where they haven't! You know where your hearing cut-off is. Boost the frequencies above that cut-off in an EQ and see if you can hear the difference in the sound file. Give an A/B test to a friend to see if they can hear the difference (an A/B test is a simple comparison between two objects where the person being tested doesn't know which one is which).

Exercise 5.9   Turn It Down!

Earlier we discussed a weird perceptual phenomenon related to our hearing : bass requires a higher amplitude to appear at the same perceptual volume as the mids (see section 1.3.1). The mid-frequency range becomes flattened out as we increase amplitude, leaving the perception that low and high frequencies are louder, or boosted in the signal. We can create the illusion of loudness by boosting the highs and lows, and leaving the mids where they are. Try it out for yourself.

5.4   Filters

Filters can work like EQs, in that they can boost or reduce (or even remove) frequencies in a set range—in fact, some EQs, called *shelves*, are essentially filters. A filter circuit behaves more like a sieve, allowing some frequencies through ("passing" them) and holding back others. Filters aren't designed to boost a signal, they only cut.

Low-pass filters are designed to allow all frequencies below a particular cut-off frequency to pass through, while preventing other frequencies from passing (figure 5.7).

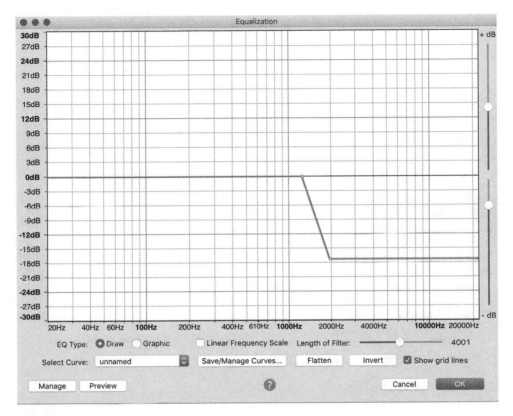

Figure 5.7
A low-pass filter, where the low frequencies are left alone and everything over 1,500 Hz has been cut.

Low pass filters have the effect of creating a "muffled" sound, much like what we might hear underwater, behind a window or door, or from a distance.

High-pass filters are the opposite of low-pass filters. They cut frequencies below a set point and allow the higher frequencies to pass through (figure 5.8).

Band-pass filters cut the highs and lows, and let a middle band of frequencies pass through. These can be useful for isolating a sound (figure 5.9).

Notch filters (sometimes called "band rejection") cut a notch out of the frequency spectrum. These can be really useful when we have an annoying sound (e.g., tape hiss, air con) and we know its frequency (figure 5.10). We used a notch filter earlier when removing the fundamental to show a phantom fundamental (section 2.5.1).

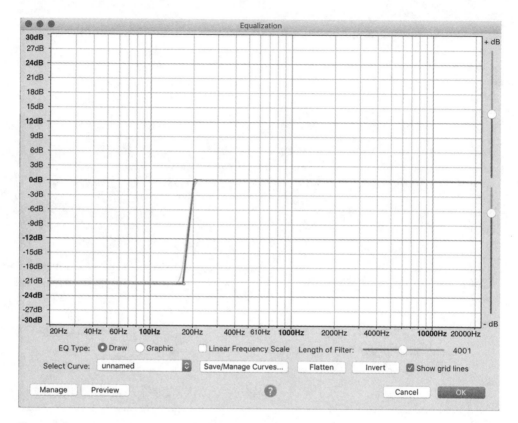

Figure 5.8
A high-pass filter, where high frequencies are left alone and everything over 200 Hz has been attenuated.

There are other types of filters, which are less commonly in use. *High- or low-shelf* filters boost or reduce frequencies in a particular range, but don't cut out frequencies. *Comb filters* are what gives the phaser its distinctive sound, but instead of sweeping the amplitude, comb filters are a harsh on-off, like the teeth of a comb. We can imitate a comb filter by making a "FFFFFFF" sound with our mouth, and putting our flat hand on and off our mouth. It has a quick on-off pattern caused by the continued interference on the signal. There isn't a built-in comb filter in Audacity so we have to download one to explore it.

Filters are sometimes used for effects (such as the comb filter), but even more often they are used to correct audio files by eliminating certain problems, such as a buzz or

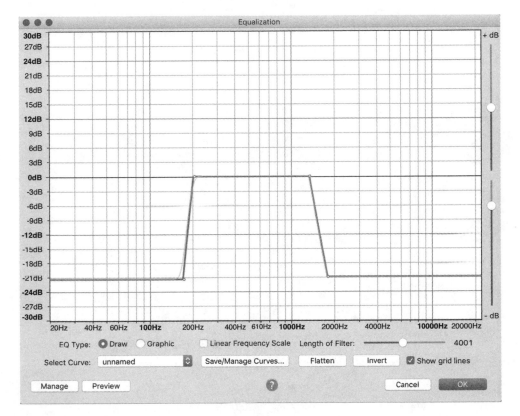

Figure 5.9
A band-pass filter, which leaves a band and cuts the highs and lows.

a hum, pops, tape hiss, or other unwanted background sound. If we can figure out the frequency of the refrigerator or air conditioner in the background on our recording, for instance, then we can filter out that frequency without messing with the other frequencies.

Exercise 5.10   Exploring Filters

Choose different filters and set different cut-off frequencies to hear the differences in how they impact the sound file. Here is where you'll find the Q value we talked about earlier: the slope of the curve.

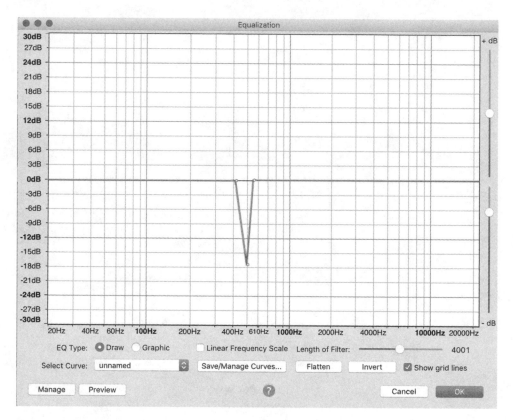

Figure 5.10
A notch filter at 500 Hz.

5.5 Modulation: Ring Modulation and Vocoder

A *ring modulator* is a type of modulation effect that creates a metallic or robotic sound. Ring modulation is created by taking two sound signals and multiplying them together to create two new frequencies that are the sum and difference of the inputs. So, if you were to put in two sine tones at 500 Hz and 600 Hz, the output would be tones at 1,100 Hz (600 + 500) and 100 Hz (600 – 500) played at the same time with the original tone—in other words, we're not just hearing the combination tones in our ears (section 2.5.1), but they're now part of the file. So if we generate a sine wave at 500 Hz and select a modulator at 600 Hz, we get a new waveform that, if we open up the plot analysis graph, we can see has generated sounds at 100 and 1,100 Hz. You'll need to download a plugin for the ring modulator. (`Generate > Tone > 500 Hz, Effects > Ring Modulator > Modulation frequency: 600 > OK`) (figure 5.11).

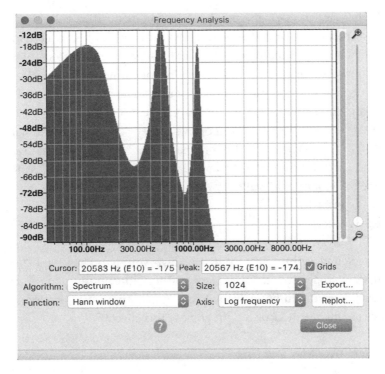

Figure 5.11
Plot Spectrum analysis of ring modulator on a 500 Hz sine wave with a 600 Hz modulator.

Usually we aren't inputting sine waves, but one sound source (known as the *modulator*), which is then mixed with a sine or square wave (known as the *carrier*) as the secondary input signal. The result in more complex sounds is that the harmonics created tend to not be even, so the result can sound metallic, hollow, and bell-like, particularly at higher frequencies, which produce more audible resonance.

Exercise 5.11   Ring Modulator

Record your voice and some other sounds, and apply ring modulation. How does the ring modulator affect the way it sounds? Go back to the section on difference tones (section 2.5.1) and create the difference tones with your ear and then with the ring modulator on a sine wave. What is the difference between what is generated by the ring modulator, and what is generated by your ears?

A *vocoder*, or voice encoder, also uses a carrier and modulating wave, and can sound a bit like a ring modulator. Vocoders work across frequency bands, so a broad frequency spectrum is required, particularly in mid to high content (which is to say, it won't really work on a sine wave, but needs a sound with lots of harmonics). Typically the modulator is a voice, but it can be used on other sounds. The modulator is split (grouped) into many bands of frequencies, and each band is sent through a filter set to the same frequency that was analyzed. The carrier wave is then sent through the same band of filters, and filters out elements in the original signal by imposing the dynamics (an envelope) and spectral content of the modulator. The effect is pitched voice (if using voice), where the input signal will follow the synthesizer signal's frequency.

Exercise 5.12   Vocoder

Grab some sound samples or record your own voice. You'll need a stereo file for applying the vocoder. If you didn't record in stereo previously, you can change this in the drop-down menu beside your microphone. Apply the vocoder. What is the difference as you hear it? How might you use this effect?

5.6   Distortion: Overdrive and Fuzz

Overdrive, fuzz, and *distortion* create an artificial distortion of the sound, effectively clipping either the fundamental or the harmonics of the sound. The effect is a result of the sound being overdriven—the sound volume is pushed too high, causing the signal to distort. The result is the addition of and/or boosting of harmonics to the sound—a little bit of distortion and it can sound "warm and fuzzy," a lot and it can sound harsh and "dirty."

Some people distinguish between overdrive, fuzz, and distortion, with overdrive typically described as a "soft clipping," or subtler effect, and distortion as "hard clipping," a crunchy sound, and fuzz as even harder or more aggressive clipping and saturation of harmonics. Often, however, these terms are just used interchangeably. Technically, overdrive and fuzz are both types of distortion, where overdrive is subtler and is "cleaner" at lower amplitudes, boosting saturation as the volume increases. EQing (boosting) the harmonics of a sound, rather than boosting the main signal, is one way to create a saturated, distorted effect without increasing the overall amplitude.

Soft clipping doesn't cut the waveform off so much, as instead it rounds it out at the cut-off (0 dB: figure 5.12), whereas hard clipping, as we saw in chapter 2 (section 2.4.2), cuts the tops and bottoms of the waveform off.

Figure 5.12
A soft clipping effect on a sine wave.

The human body overdrives its voice at times of anger or anguish: we have to yell quite hard to overdrive our voice. The resultant effect of overdrive, then, can be a perception of anger, anguish or excitement. Adding a little distortion—boosting the harmonics of the sound—can also increase the perception of loudness without actually increasing loudness, which can lead a sound to feel either warmer (if used in moderation) or angry/loud. Because a lot of harmonics will push a sound toward loud noise, the effect is used a lot in genres of heavy metal, punk, and industrial music.

5.6.1 Digital Distortion

Audacity has many settings for distortion. There are preset types that will alter the harmonics or other aspects of the distortion type. The distortion type will determine which parameters you can adjust in the window below: some sliders will be disabled for some distortion types.

Clipping level: peaks greater than the level set are clipped off.

Drive: the amount the signal is amplified prior to being clipped.

Make-up gain: the amplification of the final output.

Distortion amount: the amount/strength of distortion.

Exercise 5.13 Distortion

Try some different distortion effects and note for yourself how the sound appears to change (Effects > Distortion > Distortion Type). Change the settings for each type, as well.

5.7 Summary and Bonus Exercises

We've looked at a wide range of effects now. Knowing what effects do to a sound and the impact that has on the perception of that sound is really important, and is a staple of sound design. While many of the effects are based on analog tape-recording techniques and effects pedals that were experimental in the 1960s and '70s, today sounds are usually recorded dry and digital effects are used to adjust the sound to what we want. In this way, we can have more range and more easily adapt effects to what we want. Quite often, effects are used not in isolation but in combination with one another. To create a walkie-talkie effect, for instance, we can use a band-pass filter to filter out the highs and lows, and add distortion to "dirty up" the sound. Try to experiment with combinations of effects to mimic various effects you've heard in movies, songs, and so on.

We've now got a lot of experience applying different effects, so let's put them to creative use with some exercises!

Exercise 5.14   Guess the Effect (Partner Exercise)

Grab a partner. Use the same sound sample and manipulate it using a different effect. Swap your results, and see if you can then mimic your partner's effects by guessing what they did. If it's too easy, add a second effect on top of the first, and try again. Then a third effect, and so on.

Exercise 5.15   Change a Sound

Take a sound and put it through as many effects as you need to turn it into another sound. For instance, try to change the sound of a kitten into the sound of a dog. Try to change the sound of a running car into the breathing of a person. Log your steps in effects chosen.

Exercise 5.16   Build a Sound from Memory

Think of a sound from memory. Record or download a similar sound and then use effects to try to adjust the sound until it matches what you had in your memory. Log your steps in effects chosen.

Exercise 5.17   Glitch

"Glitch" a sound: apply some effects on a sample so that it appears to have been a glitch. A glitch might be a tape deck sound that suddenly slows down part of the track, or a digital glitch like a quick skip, aliasing, or other sounds that make it sound like a mechanical error.

Exercise 5.18   Clean Up Aisle Audio!

What happens if we have a sound effect sample that has some artifacts we don't want? We can edit it knowing what we now know about filters and effects. Find someone's really bad recording online (for instance, a podcast, a YouTube video) and try to improve their sound as much as you can. Don't forget to write down your steps taken.

Exercise 5.19   Subtraction

Take one of your field recordings. By *only subtracting* sounds from it using filters in different places, turn it into something else (you can't move or alter the sounds in any other way).

Reading and Listening Guide

Jonathan Weinel, *Inner Sound: Altered States of Consciousness in Electronic Music and Audio-Visual Media* (2018)
Weinel's book explores altered states of consciousness (drugs, dreams, trance, etc.) and how they are represented in and potentially induced by audiovisual media, with a focus on music. A fascinating read of how we virtually re-create physical sensations.

Dave Tompkins, *How to Wreck a Nice Beach: The Vocoder from World War II to Hip-Hop, The Machine Speaks* (2010)
Tompkins follows the history of the vocoder from its invention in 1928 to today. The title takes its name from a misheard vocoder recording: "How to recognize speech." It's a bit haphazard, but with some interesting historical gems and lots of photos.

Karen Collins, "Sonic Subjectivity and Auditory Perspective in Ratatouille" (2013)
In this article, I explore how a subjective perspective is created for the audience in the film *Ratatouille*. For example, in one kitchen scene, the sound designer Ben Burtt creates a subjective position for the audience by the use of low-pass filter and reverb effects when we are situated "with" the mouse under the metal colander.

Stalker (dir. Andrei Tarkovsky, 1979)
The story for this film is about an expedition led by a guide taking two clients into a mysterious restricted site known as the "zone." The scene to focus on takes place on a

railroad cart from the known, natural world into the zone. One man asks the guide if they will get caught, and is told that the police won't go in the zone because they're afraid. The question "Afraid of what?" hangs in the air for about four minutes as the visuals just cut back and forth between the two clients. The scene is carried by the sound of the rail car: as they get deeper into the zone—and as we move deeper into their mindset—an increasing amount of effects are placed on the sound, so that by the end of the scene they are saturated with phasing and filters and reverb.

Music Listening
Psychedelic drug effects were common in 1960s psychedelic rock music. For a sample, see the Spotify playlist on the studyingsound.org website. What sound effects did they use?

6 Mixing

We call the balancing of the combination of sounds *mixing*, and the mix involves the overall composition, emphasis, dynamic range, and spatial position of sounds. Mixing determines the relationship between sounds in a composition—whether it's a song, a soundscape, or the sound for other media. To use a visual analogy, sounds are composed in the mix much as we learn to compose images if we study art: we learn how to direct the attention and keep the eye moving around a composition, and we can do this with the ear and sound. We try to add clarity and direct the listener's attention to the most important elements.

Everything we've learned so far comes together in the mix. Mixing involves many aspects we've already talked about in regard to the use of effects and creating a sense of space. We've already covered many effects used in mixing like equalization, and we'll cover spatial positioning a bit more in this and the next chapter. Here we're going to focus on a few other concepts related to putting sounds together. Mixing is usually done by a mixing engineer: someone who specializes in the task. It's an enormous area, and takes a lot of practice to get right. There is a lot more to know about mixing than what we can cover here, but we'll tackle the basics and get you started.

6.1 Mixing Theory: Three-Dimensional Sound

We looked at how sound responds differently in space in chapter 4, and we'll continue this work in chapter 7. But another aspect of sound in space relates to mixing, namely, the perception of sound as part of a composition (whether music, film, soundscape, etc.) that is part of a three-dimensional space, even when we mix in stereo. With two ears, we hear sound as existing in space around us. This perception is enhanced by a good mix, which typically situates us in the center of this sonic space. I find that it's useful when mixing to think of sounds as occurring in space in terms of four dimensions:

three physical dimensions—in visual coordinates we speak of these in terms of the Cartesian x/y/z co-ordinates, which consist of left-to-right (usually x), top-to-bottom (usually y), and near-to-far (usually z) (figure 6.1). The fourth dimension is time: how things change over time is an important aspect of the mix, but let's focus on the first three dimensions for now.

The left-to-right or x axis is controlled by our panning settings. We will discuss panning in further detail below and in the next chapter; the first step in understanding our "sound box" (see Moore 1992) is the left-to-right position. This position is set virtually with panning, but there are also perceptual aspects to panning that can alter where we think we are putting a sound in the field, and this perception shifts depending on where the listener is located with regard to speakers, and on whether they are using headphones. Panning algorithms use a complex mix of effects to create the sense of distance, including:

1. Volume (sounds will appear louder as they move toward the center)
2. Spectral adjustments (filters and equalization)
3. Phase changes
4. Timing and delays
5. Reverberation

The height or y axis is created through the perception of frequency: high frequencies are perceived as being physically higher in space, and low frequencies are perceived as physically lower in space. In part, this can be attributed to the bioacoustics of the human body: higher-frequency sounds tend to resonate higher in our body (e.g., nasal cavities) and lower-frequency sounds tend to resonate lower in our bodies.

The z axis or the near-far axis is created perceptually through a number of phenomena:

1. Louder sounds appear closer to us.
2. Sounds without reverb (when used in conjunction with sounds with reverb) will sound closer to us (since the reverberation implies a space between us and the sounds).
3. Sounds that have been closely miked will appear closer to us.
4. Sounds with unusual effects tend to stick out and appear closer to us.
5. Sounds with hard compression (see below) will appear closer to us.
6. Sounds with more high-frequency components will appear closer to us.
7. Discrete sounds against an ambient bed will tend to appear closer to us (e.g., the back-up beepers on a truck against the white noise of a construction site).

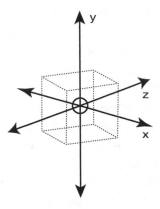

Figure 6.1
The three dimensions (x/y/z) of mixing.

We can use this knowledge to place sounds into a perceptual three-dimensional location in space around the listener, and it is this space that can form the basis of our
understanding of the mix. Mixing, in other words, goes beyond just achieving clarity
between sounds: it is a delicate balance of space over time.

6.2   A Note on Mixing in Audacity

Audacity isn't really designed for mixing, but we can create basic mixes in the program.
As you progress, you'll want to switch to a program designed for more advanced multitrack mixing, like Nuendo, Reaper, Logic, or ProTools. So, while you *can* mix in Audacity, I recommend you invest in a professional audio software program once you've
learned the basics. Reaper is becoming increasingly popular since it offers a very affordable price for those just starting out, and offers a sixty-day free trial. We'll stick with
Audacity for now, because we can do the basics and the concepts will carry over to
other software you want to try.

You may have noticed that when we select `Mix and Render` in Audacity the amplitude increases. Mixing in Audacity adds the two waveforms together, so we will in most
cases have a higher amplitude in the resulting track. If we are mixing a lot of tracks
together, the result will be much louder than we probably intended. It's important,
therefore, to reduce the overall gain in advance of using `Mix and Render`. If we generate two sine waves with the amplitude set to 0.5 for each, and open the mixer board
(`View > Mixer Board`), we can adjust the gain on these two files separately before
we mix them (figure 6.2). If we click `Mix and Render` without adjusting the gain, we

can see the output gain is a much louder file—it's twice 0.5 (or 1) which means we're just about clipping. When we use `Mix and Render`, we'll now only have one track on the mixer board, but we can adjust the gain back down, which will alter the playback amplitude, or we can use `Effects > Amplify` and reduce the amplitude of the track itself (do this before it clips!). Note that we can change the amplitude settings in Audacity from a –1 to 1 scale to decibels, so we can be more accurate and detailed in the numbers involved. To *double the amplitude* of a track, increase it by 6 dB. To cut the amplitude in half, decrease it by 6 dB. We may want to use `Mix and Render to New Track`, which will preserve our original tracks and allow us to make adjustments on individual tracks if we don't like the render—just delete the render and try again. It's always best to keep a backup, though, in case we end up not liking the mix later. As we deal with multiple tracks in Audacity, as seen earlier, we can also mute tracks using the mute button, or solo a track if we want to focus on just one track in a multitrack session.

Mixing can involve the addition of all the types of effects discussed in previous chapters—especially EQing and filters. In addition, there are other effects that we haven't yet discussed that often fall under mixing's domain. First, however, we need to learn a bit about dynamic range.

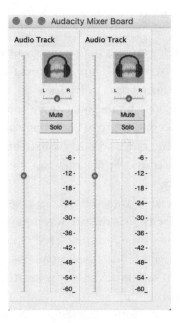

Figure 6.2
The Mixer Board window (`View > Mixer Board`).

6.3 Dynamic Range

Dynamic range is the distance between the peaks (loudness) and valleys (quiet) of a sound file over time. We can think of dynamic range like a black–and-white photograph: if the file were all black, or all white, the image would be boring. Variations—all the shades of gray—are what make it interesting. Imagine if you held a party at your house every day. Eventually, the party would be boring, because it's just a party all the time. You need down time in between the parties, the boredom of everyday life, in order to be excited about the party. By using contrast between the quiet and loud, we create interest and focus and can draw attention to particular sounds. It's important to think not just of the loud sounds, but also the quiet, and how we can use quiet or even silence to emphasize important points when we bring the loud back in. This is the role of dynamic range, the light and dark of a soundscape.

Nathan McCree, who created the sound and music for the original *Tomb Raider* game, explained:

> And in the same way that a piece of music tells a story, a soundscape can tell a story. If we take it—like a horror genre, for instance: You know you will lead the player into a false sense of security, and then just before you scare them, you will just drop all the sounds out, until there's this eerie moment of nothing. And then, "Dah!" Suddenly out jumps the monster and you jump out of your seat. So this sort of storytelling approach is the same for music as it is for sound. (quoted in Collins 2016, 20)

If we look at the four tracks in Audacity in figure 6.3, which track has the highest dynamic range? Which track has the lowest dynamic range? If you guessed track 3 for the highest, and track 1 for the lowest, you are correct. Track 1 has no dynamic range—the entire track is at the same volume. Track 3 has the greatest peaks and valleys. Dynamic range is not the amplitude, but the *changes* in amplitude over time.

We use dynamic range not just for technically representing the loud sounds, but also for emphasis and to draw the listener's attention. In a story, for instance, the loudest sounds are probably going to be in the most dramatic scenes, at the point where the audience is supposed to feel the most tension; they might not be what should be the loudest sounds in real life. But dynamic range can also be tricky, because if there is too much distance between the loud sounds and the quiet sounds, we may find our listener has to "ride the remote" and keep turning up the volume when it's just dialogue and turn down the volume when there are a lot of action sounds. Films are usually mixed for theater listening, where they're not worried about waking the baby or annoying your neighbors, so there tends to be a lot of dynamic range. Music used to display greater dynamic range, too, but over time the range has gradually gotten *compressed* (see below).

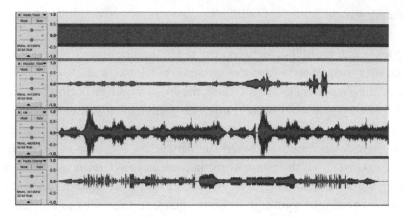

Figure 6.3
Four tracks of varying dynamic range.

Exercise 6.1   Dynamic Range

Think about the dynamic range of the listening exercises we've been doing so far: in which location have you found the most dynamic range? For your next listening exercise, focus on the dynamic range of the soundscape you're listening to. What is the loudest sound, and what is the quietest sound you can hear? What is the most important sound, and what is the least important sound, and how would you adjust the dynamic range to reflect that?

6.4   Compression, Limiting, and Normalization

The function of a *compressor* is to decrease the dynamic range of an input sound source, that is, to reduce (compress!) the difference between the quietest and the loudest sounds. The end result is to effectively make the loud sounds quieter and the quiet sounds louder. For this reason, it doesn't make much sense to compress a single one-shot sound. Rather, we would compress an overall music track, field recording, or composition. Compressing a file can be useful when parts are too soft or too loud, like a dialogue recording where someone is speaking at different volumes. The downside is, if the quiet sounds contain unwanted noise (the hum of an air conditioner, for instance), that noise may get boosted in the output, so it's best to filter unwanted sounds out first.

A compressor works by weakening the input signal only when it is above a certain set threshold (volume) value. Above that threshold, a change in the input level produces a smaller change in the output level. So as the overall signal of the file is boosted, if the signal above the threshold is too loud, the compressor brings the amplitude down above that threshold. The compression *ratio* is the change in output levels from the given input, so a 2:1 ratio means that a 2 dB change in input results in only a 1 dB change in output.

Compressors cap the sound at an absolute, which is useful when we know the desired peak amplitude of what we're mixing something (e.g., film is often compressed at –6 dBFS). We can also use compressors to extend the sustain of sounds by boosting the reverb, but this can sound flat, loud, and distorted if used too much. Compressors can be used in this way to increase the perception of loudness—a fact that some audiophiles have complained about when it comes to recent music, which in some instances has been so heavily compressed that the track seems louder than other songs or media. It's quite commonly used in television or radio commercials, which usually have a set loudness limit, but the use of compressors will increase the perception of overall amplitude.

Multiband compressors work in separate bands of the frequency input, so a full-band compressor sets the overall compression based on the low-frequency sounds. Essentially, the compressor first filters the sound into bands, and each band is then fed into its own compressor and combined back into the mix.

Normalization, another common effect, sets the loudest peak at a particular point, raising or lowering the entire file to that level and adjusting the gain by a constant value across the entire file. So with compression, we are adjusting the portion of the audio that is below a certain threshold. With normalization, we are setting the overall volume to a standard level. *Peak normalization* finds the peak volume in the file, and then uses that peak to boost the file to the set point. *RMS* (root means square) normalization finds the *average amplitude* of the file and uses that to adjust the entire file, which can lead to clipping. Normalization may be applied to batches of files, so if we are building some online podcasts, for instance, we can normalize all our files so the listener doesn't have to jump back and forth to adjust volume when continuing to the next file.

If the compression ratio is very high—a ratio of 10:1 to infinity:1, this is known as *limiting*, which is like a shelf above which there will be no increase in volume at all. Limiters are typically set only to round out high peaks, in cases where sounds must not go above a particular threshold (as in film, for instance). In digital recording, we know the maximum set loudness is 0 dBFS, so all sounds must drop below that threshold. *Limiters* are similar to low-pass filters, but affect volume instead of frequency, so the threshold will be the cut-off volume and everything above that threshold will be pushed down into the mix, while anything below that volume will "pass through."

6.4.1 Digital Compression

We'll need a fairly long (ideally, several minutes) file with a lot of dynamic range to work with digital compression. Import the file into Audacity and then use a compressor. Try different settings and listen and look at how it changes the file. The main settings are as follows:

Noise floor adjusts the gain on audio below this level so that it's not amplified. For example, too much background ambient room noise would be boosted by compressors, but if we adjust the noise floor above the room ambience, it won't be boosted.

Attack time determines how quickly the amplitude is reduced once the input exceeds the threshold—you can think of it as how quickly the compressor will respond. If the attack time is set too slow, then a short burst of louder sound can get through, which can lead to distortion. You may want to leave a slow attack sound in some cases to ensure that the full attack of a sound comes through.

Release time is the time it takes for the compressor to return to its full gain after reaching the threshold. If it is too fast, you get a pumping, breathy quality where you can hear the volume go up and down.

Make-up gain (output volume) lets you bring the compressed audio back up to an acceptable level, given that the compressor will tend to reduce the overall volume of the file. The make-up gain will take it back to a certain level, saving you a step in processing.

Exercise 6.2   Compression

Try compression, normalization, and limiting as audio effects. These effects can be used, for instance, to enhance sustain/reverb on files, and set the limit where it starts to distort. Take a lot of time on this one—it's a more delicate effect than some we've heard so far, and it takes practice to do it right.

6.4.2   LUFS: Finalizing your Mix

If you are producing content for broadcast or distribution you may come across LUFS, or Loudness Units relative to Full Scale. LUFS is a standard for measuring loudness designed to enable normalization across multiple different content sources—think of a radio that has a radio drama, with advertisements in it, followed by news broadcasts,

Figure 6.4
A file before any effects are used.

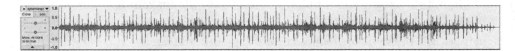

Figure 6.5
Compression with a 2:1 ratio.

Figure 6.6
Compression with a 7:1 ratio.

Figure 6.7
Hard limiting to –10 dB.

and so on. You don't want your listener to have to keep getting up to turn the volume up or down. By standardizing the loudness levels, it makes it much easier for your end user. A LUFS meter will tell you the averaged loudness of your audio file—an overall level of your file. If you had a file with a lot of loud sounds and little dynamic range, it will get played back at a lower volume, since it has a higher overall loudness than a file that has more dynamic range. LUFS loudness meters are included in Adobe Audition, but you'll need to download a separate plugin for Audacity.

LUFS standards vary, depending on where you are distributing your content and if your content is distributed in mono or stereo. The European Union has a standard called EBU R128, which recommends –23 LUFS for television. Audio for podcasts on iTunes is supposed to be set to –16 LUFS; YouTube and Spotify aim for –14 LUFS.

6.5   Expansion and Gating

An *expander* (sometimes called a "stereo image expander" or "stereo widener") is the opposite of a compressor: it expands the dynamic range by reducing the level of signals below a threshold, making quiet sounds quieter. As with compressors, the ratio can be adjusted. An extreme ratio results in a gating effect. There is no expander plugin for Audacity, but we can download a plugin called Noise Gate for gating.

Gating is a kind of inverted limiter, or a particularly harsh expander. With *gates*, all signals below a threshold are reduced by a set amount (the range). In other words, the input signal is scanned for sounds below a certain threshold, and if they are below that threshold, the gate closes and the sound is muted, so if we have a hum in the background, we can get rid of it by reducing the amplitude of sounds that are in that range.

Gates are used to "add punch" to percussion in music by shortening the decay time, or to cut the level of noise between sections of a recording.

Exercise 6.3   Gating

Make a recording with some background ambience behind some discrete sounds. Try different gating settings and see if you can get rid of the ambience.

6.6   Ducking

Ducking reduces all signals above a threshold by a set amount (the range). Ducking is used in dialogue in film or radio when the music must get quieter so that we can hear the speaker's voice. The "background" sound is "ducked" under the sound we want to hear. For instance, if you have an overdub of a foreign language, you may hear the original language ducked underneath.

Audacity comes with an Auto-Duck plugin. We need a *control*, which is the signal we want to keep (e.g., the voice). The control should be on the bottom in track order. Then we have the signal that is the "background"; in this case I generated white noise. Select the white noise and apply the Auto-Duck. The background signal will now duck under the voice. You can see (figure 6.8) where I've paused in the voice track, the background ramps back up. Note that when we're dealing with multiple tracks, we can change the track order by clicking on the drop-down menu where we changed between Waveform and Spectrogram view (or the track name): here we have the option to move the track up or down. Incidentally, we can also change the track color here, which can be useful when working with multiple tracks.

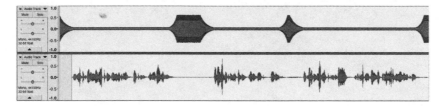

Figure 6.8
Ducking in Audacity. The top track is the music that is ducking under the dialogue track (track 2) files.

Exercise 6.4   Ducking

Try some different ducking settings. How might you use this as an effect on something other than voice?

One term you'll come across with effects is *sidechaining*. A sidechain feeds a different signal to an effect, so ducking is an example of side-chaining: it's got a control file that tells the original file where to drop down. So if we wanted to apply an effect to a file, we feed it data from a second file instead of the original file. This is probably most commonly used in compression effects, where, for instance, the kick drum operates in a similar frequency to a bass synth in a dance track. A compressor is used on the bass every time the kick drum occurs, to give the kick some "oomph" and avoid the "muddy" sound created by having two tones in the same frequency band. But side-chaining doesn't have to be used for practical purposes—we can use sidechaining on sounds with all kinds of effects for creative purposes. You probably won't be able to find these tools for Audacity, but they are common in other DAWs.

6.7   Noise Reduction

If we have a noisy track, most audio editors now have built-in noise removal plugins. To remove the noise, we have to teach the program what to remove by selecting a sample (control) of the noise as a *noise profile*. Using the mixed noise we created in the last example, select `Noise Reduction` from effects, highlight a few seconds of noise and get the profile. The profile is now stored and we have to select `Noise Reduction` again. Now we can reduce the noise by a set dB, set the sensitivity of the reduction, and apply it to the file. Listen to the difference between this noise reduction and ducking. The trouble with noise reduction is that it's going to remove some of the frequencies we

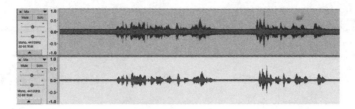

Figure 6.9
Before and after noise reduction.

want to keep—in this case, some frequencies from my voice, so the voice sounds a bit unnatural. There are ways we can tweak a bad recording, but as the saying goes, garbage in, garbage out! It's always better to take the time to get a clean recording to begin with.

Exercise 6.5   Noise Reduction

Find or record a noisy file, and try to reduce the noise. Try files with varying levels of noise (you can find some with low ratings on Freesound for a starting point). See if you can find the balance between reducing noise and not interfering with the main signal.

6.8   Figure and Ground: Signal to Noise

We talked earlier about the signal-to-noise ratio (section 2.8). Understanding the right signal-to-noise ratio for our audience is an important part of mixing. Some people like to think in terms of figure and ground, which is used in the visual arts to describe the subject and the background. We can also talk in terms of figure, field, and ground. Some sound designers refer to these as the foreground, midground, and background, terms also commonly used by cinematographers when they talk about image in film. We can also think of these as the focus, support, and background.

The *foreground* is what we want listeners to focus on—it should be the most active part of their listening. Speech is usually the most foregrounded element in film, for instance, or the voice or lead in music. The *midground* contains sounds that are there to support the foreground image. They may be Foley sounds, or sounds that are part of the scene to be noticed but not focused on. They are often isolated, one-shot sounds or repeating sounds that are part of the scene. The *background* is the ambient bed that is part of the setting of the stage.

If we think of a theater set, the foreground would be the actors, the mid-ground would be the objects they pick up, use, or move around, and the background would be the scenery. In sound, then, the foreground would be the actors' voices and anything they do with objects that is really important (e.g., slapping someone with a pair of gloves); the midground would be sounds made by objects that are part of the scene but not necessarily important, like a chair moving or footsteps; and the background would be the ambient effects that tell us where they are, like wind or rain. Another example is an emergency services scene. An ambulance siren might be a foreground sound, immediately telling us there is an emergency. Midground sounds might be people yelling, a fire burning, and so on. And ambience might be nighttime sounds around the scene, road traffic near the scene, and the like. What sounds constitute foreground, mid-ground, and background depends on the context and what we want the audience to focus on.

In section 6.1 we talked about the perception of near and far in the mix. These are the parameters we can play with to get the right perceptual distance between sonic objects in a mix. By foregrounding sounds in a mix, we focus the audience's attention on those sounds and draw their attention to them. Thinking about what is most important to the audience is the first step in thinking about our mix: What is the most important sound they hear? What is the least important sound?

Exercise 6.6   Grouping: Figure and Ground

For your listening exercise today, group the sounds you've described into foreground, midground, and background sounds. Then record a single sound as a foreground, midground, and background by increasing your distance (and potentially mic axis, microphone polar pattern, etc.). Play the recordings back: how does the shift in perspective change the way you think about the sound?

Exercise 6.7   Too Much Noise

How many sounds can you layer into a mix before it gets to be too much and just becomes "muddy"? How can you adjust that same number of sounds to separate them out further in the mix?

6.9   Panning

Panning involves the perceptual position in the stereo field (as defined above, the x-axis). As we know, stereo files have two tracks, and there's usually something different in each one. Each track is assigned to one channel and is designed to be played through one of two loudspeakers or headphone sides. Typically, the loudspeakers in stereo are set at 60° from the listener's head. We call the head's position in this equilateral triangle the *sweet spot*, since it gets an even mix of both speakers.

Professional audio software often has advanced stereo-imaging plugins, which can give you a sense of the exact placement of sounds in the stereo field. Izotope makes a free stereo imager called Ozone that you can download from their website. The Waves S1 Stereo Imager, which is not a free plugin, allows you to change visually not just the spread but the rotation and symmetry of the stereo field. Stereo imagers allow us to get a better sense of where we're placing the sound in the stereo field.

Panning algorithms don't all work the same way; some vary in how they adjust the amplitude of the sound signal as we move further from center, so it's important to understand how we're adjusting our sound. It is not just assigning a sound to a channel (left or right), but perceptually placing that sound nearer to or farther from the center through volume, timing adjustment reverberation, and filters or equalization. The perceptual distance from our ears, then, which we'll talk about further in

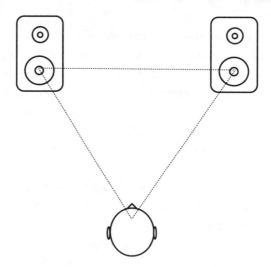

Figure 6.10
Optimal speaker setup for listening in stereo.

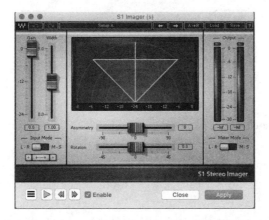

Figure 6.11
Waves S1 Stereo Imager.

chapter 7, is a complex process, and advanced virtual panners can really do a lot to change the perceptual location of sound.

In Audacity, we're limited to very simple panning settings unless we download extra plugins. The panning in Audacity can be done in a few ways: the entire track can be panned using `Tracks > Pan` (although here we are limited to left/right/center). We can also use the pan slider in the track control panel. If we hold down the shift key, we are given smaller increments to adjust the slider on a finer level; otherwise, we are given 10 percent increments.

While we don't have variable panning within a single file in Audacity, we can appear to adjust pans by layering the same sound and using cross-fades while switching between left-panned and right-panned files. We can also put different pans on different files within the mix before we do the final `Mix and Render`. By varying the apparent distance from the center, we can separate sounds from each other and make them more interesting.

Exercise 6.8   Panning Perception

Set two speakers at a 60° angle to your head. Pan a sound all the way left and another all the way right. Where does the sound appear to come from? Most people find that sounds appear farther away than the speaker. Where does the sound appear to come from in terms of the front–back location, and how does that change as you change the volume? What about height and frequency?

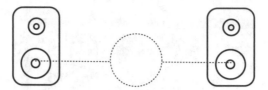

Figure 6.12
A phantom center created by two evenly panned sounds.

6.9.1   A Phantom Image

When we have two or more speakers, we can create the perception that the sound is coming from somewhere else with panning. With stereo speakers, a phantom center is *also* common if the speakers are playing the same sound at the same volume, and the listener is in the sweet spot (figure 6.12). If the listener moves out of the sweet spot, the phantom image can collapse. The phantom image also has a different coloration (different spectral quality) than if a real speaker were located in that spot. Panning sounds alters the phantom image.

> **Exercise 6.9   Phantom Imaging**
>
> Try to position a sound in the phantom center position using stereo. Now play with the pan on those speakers and listen for the perceptual changes. What happens when we start panning?

6.10   Mixing across Media Devices

One of the challenges of mixing is the fact that our listeners may hear the mix in circumstances very different from a nice mixing room with speakers positioned correctly and the room acoustics optimized for hearing sound. We might spend many days getting a mix to sound perfect in the studio, but when we take it to our car or play it on headphones, it sounds terrible. It's important to try the mix out on different media and get a sense of how different formats influence the overall mix. Scott Gershin, a sound designer and mixer for Hollywood and for games, explained to me some of the challenges of mixing:

> It's fun when you get to do multichannel with subs and you get to rock the room, and you get to be the rock star with sound, whether it's movies or games. . . . But what's happened is we're expanding the market. It's not just that. You know, we're in a time, an era, of disposable

entertainment and entertainment on demand. So at that point, people will go for a little lesser quality—think about the music we're listening to, and your MP3 is not a great-sounding format, but it's convenient. Neither was the cassette, but it was convenient. So people like convenience, and they're willing to trade off.

So the question, then, for us becomes, how do we bring the best experiences? When I started in this business we were 8-bit. It was really horrid, actually. We started with very low sample rates; pick your sixty best sounds, at best. So when I see mobile, I think, "Hey, we're kind of back here again." A little bit better audio quality, but we're kind of back here. You know, at that point, you have to be able to make those decisions—how to best support the story or the game or whatever you're working on within the environment you're at.

There are a lot of different tricks—compression tricks, the way you approach a mix is very different, different dynamics if they're wearing headphones. Are they going to wear headphones in a noisy environment? Is there dialog involved? How's the music going to be involved? So at that point it creates a different kind of technical challenge. So you then have to figure out what medium you're being played back on. And then you create your technology, your techniques, your trade, your artistry, to fit that environment. . . . But for me that's kind of what makes it fun, in that it's not easy, you've got to figure out maybe a different way of working, and one thing I love about the gaming world that's a little different from film. (quoted in Collins 2016, 184)

6.11   Technical versus Creative Mixing

In addition to thinking about different listening formats, it's also important to think about mixing as an art form. It is perhaps tempting to get as close to the "real world" as possible in a sound design mix. But this is not always the best choice for our creative projects. There are plenty of examples of mixing where there is an intentional intervention into the "real world" approach, instead applying more creative uses of the mix. The films of Jacques Tati are an excellent example: not only do the sound editors on the film make some unusual choices in terms of the sounds chosen for events, but the sounds are often mixed at unusually high volumes. Rather than being purely technical, many creative choices need to be made in the mix.

Exercise 6.10   Technical and Creative Mixes

Take a group of samples to create a soundscape. Try to make it as realistic as possible. Then, play with the mix to illustrate the following concepts: The person in the scene has just been hit hard on the head. The person in the scene is raging. It's a warm, fuzzy flashback. It's all in someone's head.

6.12   Point of Audition: Objects in Ears May Be Closer Than They Appear

When we mix, as described above, it's important to think about our listeners and what they need to hear. We can create a *subjective position* for our listener—we can position them as "in" a space or character, or external to that space. We call this the *point of audition*, roughly analogous to point of view. Decades of film research have shown how camera position can alter subjective experiences of scenes, whether we're put in the place of a character, or an inanimate object in the scene. Think, for instance, of how the positioning of a serial killer watching his prey is often shot with us in the killer's point of view (POV); alternatively, we can be positioned in a POV shot as the victim, looking into out into the dark, afraid. By changing POV, we can be positioned as inside and involved in a scene, or as third-party observer. In film and TV, the audio mix often mimics the angle and distance of the camera. Although in many cases the auditory and visual perspectives don't match, usually the camera and sound reinforce each other, so if the camera is close, the sound is usually close too.

There are several techniques to accomplish this point of audition:

(1) *Emphasis in the mix through panning, volume, and equalization*: The most obvious way to position the audience is through the position in the sound box described above. As shown, it's not as simple as loud = close, quiet = far. There are details in the sounds that we hear if items are closer, and these need to be emphasized in the mix as well. The use of equalization can also help to emphasize some aspects of sounds over others: bright sounds that stand out and are in our most important frequency range (i.e., the speech range) can be emphasized even without altering their volume.

(2) *Filters*: In addition to the volume and EQ, we also know that sound frequencies drop off as they travel over distance or meet surfaces that absorb some frequencies. The use of filters to filter out some frequencies can help to create a sense of distance or place. As we saw earlier, a low-pass filter can create the illusion of sound obstructed by a window or wall.

(3) *Effects*: Reverberation will create a sense of space and of distance between the listener and the sound. When we looked at reflections, we determined how far we need to be from a sound to get more reverb than direct signal. If there is a lot of reverb, we know we are some distance from the sound source. Likewise, more early reflections can lead to the perception of a smaller, more claustrophobic space. But we can also use reverb, phasing, or other effects to position sounds as coming from inside someone's head by contrasting those sounds with the natural world around

them. We can, in other words, create a psychological mix that places us in the minds of characters—as we saw when discussing the rail cart scene in Tarkovsky's *Stalker*. The use of unusual effects can also make a sound perceptually closer to us, since we're likely to pay attention to something more unfamiliar.

(4) *Microphone selection*: We know that some microphones capture more detail and boost certain frequencies. The selection of microphone, then, can influence the perception of distance. Some microphones are better at capturing more delicate sounds as well as different frequencies, and some have a faster transient response time (the ability to respond to changing sounds), which influences the timbre of the sound. For instance, a condenser microphone such as the Neumann U87 has a nearly flat frequency response except for a slight lift in frequencies from about 5k Hz to 15 kHz. This higher frequency range will catch more sibilance on a voice, for example, and give the impression of closeness.

(5) *Microphone location*: Where we place our microphone is going to make a big difference in the perception of position. We looked at close-miking and proxemics earlier, and we know that placing the microphone further away from a source changes both the amount of environmental noise and reverberation that it picks up, as well as changes how much detail we get from the higher frequency components of the sound. A closer microphone will capture small sounds that even our ears may miss. On vocals, this closeness means capturing mouth smacks, tiny clicks from inside the mouth from tongue and teeth. The angle can affect the "warmth" of a sound, as well as the nasality of the noise when capturing vocals. Also, when the microphone is close to the object, it causes what is known as the "proximity effect," which means a boost or lift in the lower frequencies of the sound, making for a richer, "fatter" effect. The mix of direct (no reverberation from the room) and indirect (reflections) sound can influence the perceived distance (close-miking will have stronger direct than indirect sound).

(6) *Channel/speaker location*: We may have two or five or twenty speakers to position sounds in. How far the physical speakers are from us will influence our perceptual distance. We will hear more detail on headphones or close speakers than we will from speakers that are farther away from us. We will also pay more attention to sounds located in the front speakers (and in particular the center, or phantom center) than we will in the surrounds.

(7) *Spatial location of the sounds*: We'll take a look in the next chapter at spatial sound, but in brief, headphones can lead to in-head localization, and we can even use that as a device to place sounds into a listener's head. Where we position sounds

virtually in the spatial field, then, can emphasize or deemphasize their position and help to create a perspective. Sounds positioned outside the 60° angle location between our head and speakers can sound farther away.

Exercise 6.11   Point of Audition Analysis

Listen to a few different sound mixes—from music or film or games. Think about where it positions you as the listener. What impact does that positioning have on your subjective experience of the mix? How loud are sounds in relation to each other? What frequency bands do sounds appear in? Does the mix stay the same as the track progresses? How long is the reverb on each sound—is it one reverb on the whole file, or are different sounds treated differently? How have sounds been placed in the stereo field?

Exercise 6.12   Play with a Mix

Make a soundscape and play with different aspects of the mix to change the subjective position: record at different distances, and with different mics, and explore how much you can then play with those recordings in the mix to alter the subjective position. Attempt to create perceptions of near, far, and inside a character's head with the same sounds.

Exercise 6.13   Play with a Mix 2

Record/find all the sounds you'd need to create a car crash scene. Mix it as if you are outside the car, then mix it again as if you are inside the car. What changes did you need to make and why?

6.13   Summary and Further Mixing Exercises

When we've put so much time into our design, it can be tempting to rush a mix to get it done and move on to the next project. As shown, though, mixing can take a lot of time to get just right, and can have a big impact on how your audience hears your work. It's important to take your time on mixing. This means leaving enough time at the end of your project to do a mix, then leave it and come back to it after you've had a good break, like a night's sleep, to listen with fresh ears. When your ears are fresh, you'll hear all kinds of things you missed on the first few passes of your mix.

Exercise 6.14   Reference Mixes

When you listen to movies, music, podcasts, radio, or other media, focus on their mixing. If you like the mix, be sure to record it so you can use it as a reference for your own work later. What were the differences in the mixes you liked, and how does that change the feel of the sound? Find a mix that you don't like. What do you think is wrong with it, and how would you change it?

Exercise 6.15   Stacking Sounds

Mix two sounds together by stacking the layers to create a new sound. Leave the sounds dry (don't use any effects). Then use three sounds. Then four sounds. At what point do you just get a muddy mess?

Exercise 6.16   Stacking Sounds 2

Repeat the previous exercise, but with effects this time—what effects can you put on multiple sounds to help to reduce the mud while retaining the sounds' uniqueness?

Exercise 6.17   Take Your Mix Out into the World

Now that you've got some mixes, try them out in different settings: your iPod, your car, your friend's stereo, and so on. How does the sound change from place to place? How would you balance the trade-offs necessary if you knew most people were going to listen to your mix on a $10 pair of smartphone headphones over Bluetooth?

Reading and Listening Guide

Walter Murch, "Dense Clarity-Clear Density" (2005)

A two-part article written by Murch about the womb tone and listening as a sense in the first part, and then in the second part, an important and influential treatise on mixing. Murch takes us through his theories on mixing and then presents a breakdown of his mixing approach on *Apocalypse Now*.

David Gibson, *The Art of Mixing: A Visual Guide to Recording* **(2018)**
Although Gibson's book is about mixing music (as are most other books on mixing out there!), the techniques and examples are useful to practice listening. Gibson breaks down popular songs in terms of graphic visual shapes and colors that illustrate their position in the mix—the basic concept of the 3D sound box I drew on here. Reading the separate mix breakdowns and listening to the songs is a great way to help tune your ears to the mixer's job.

A Quiet Place **(dir. John Krasinski, 2018)**
This film is a sound designer's dream job in many ways: sound plays a pivotal role, since the aliens detect people through sound. But it's also a great example of changing the subjective position. The character Regan is hearing impaired, and the audio cuts back and forth to help us experience many scenes from her perspective. Watch the film, then watch it again with headphones, and really focus on the effects used to change the subjective perspective.

7 Surround and Spatial Sound

We've been dealing with standard stereo presentation up to now, but there are other ways we can create the spatial perception of sound. As with all other aspects we've been looking at, spatial audio is an enormous area of practice and research, and it is the one area of sound that is significantly in flux at the moment. Although Audacity has a rudimentary surround sound option, it has no native spatial sound tools. The skills and knowledge are quite specialized, but an overview to help you understand the area is useful. To do any real work in spatial audio, though, you'll need to switch to one of the other DAW tools available. We also need to distinguish between *surround sound*, which is the presentation of sound on multiple speakers (the hardware), and *spatial sound*, which is the location of the sound in a 3D space, accomplished in the software.

One of the goals of surround sound is immersion and *envelopment*. Envelopment is the sensation of being surrounded by sound or the feeling of being inside a physical space (enveloped by that sound). Most commonly, this feeling is accomplished through the use of the *subwoofer* and bass frequencies, which create a physical, tangible presence for sound in a space. A subwoofer is specially designed to handle the lowest frequencies in the human range: from 20 Hz to about 200 Hz (the top end of the frequency range varies according to specification, with 100 Hz common for professional live sound and 80 Hz for THX certification, but 200 Hz for consumer-level products).

In film, the sense of envelopment is perhaps the main reason for using surround sound, but when it comes to some other media like games, another really important purpose to spatializing sound emerges: to *localize* sound (that is, to locate sounds in space).

7.1 Human Sound Localization

Humans have two ears, and this simple fact means that we are quite good at localizing sounds. There are three components to the sound's location in space:

(1) *Azimuth*: the left–right position on one horizontal plane. When we adjust sounds in a stereo mix by panning, we're adjusting the azimuth, since the speakers in stereo are located at the same height.

(2) *Elevation*: the height of the sound, or up–down. Although we perceive high frequency sounds as physically higher, in this case we're talking about the actual location of the sound in space (for instance, played back through speaker at height).

(3) *Distance*: the position from the listener, or near–far. Amplitude can alter the perspective of nearness and farness, but as we saw in discussions of microphone positioning, perceptual distance is complex. Here we're talking about real distance from the listener to the source.

Sounds located in front of us are the easiest for us to localize; sounds behind us (and to a lesser extent to the sides) are harder to localize accurately. The reason for this, we learned in chapter 1, is that the pinna is, in part, responsible for our localization ability. Sounds coming from behind do not reflect in the pinnae and so are harder for us to get a precise localization. But the sound itself can influence how well we can determine

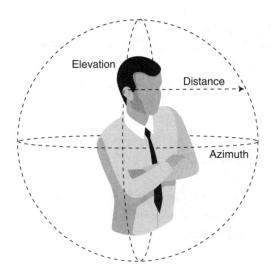

Figure 7.1
Human sound localization.

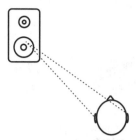

Figure 7.2
Time difference and level difference between our two ears.

where it comes from. We are quite poor at localizing bass frequencies and sounds without many spectral properties (sine waves). Sounds with close frequencies can sound like they are coming from the same source, which is why ambulance and police sirens are made up of multiple tones—we need to be able to localize the sound quickly, so by altering the frequencies we can more effectively locate the source.

Because we have two ears, there are differences in how the sound signal reaches those two ears, which we use for localization (figure 7.2). The *interaural level difference* (ILD) refers to the fact that sounds experience a slight reduction in amplitude (sound pressure level) from the sound to the farther ear. If the sound is directly in front or directly in behind the listener, the cue will be the same in each ear, making it harder to localize the sound. The *interaural time difference* (ITD) refers to the fact that sounds take a little longer to reach the farther ear.

At times, sounds can have the same interaural time difference and the same interaural level distance; this can result in what is called the *cone of confusion*. For instance, in figure 7.3, two sounds, B and D, occur at the same elevation and distance, but at a different azimuth: they will reach the ear at the same time. Likewise, sounds A and C are on the same azimuth and distance but at different elevations: they will reach the ear at the same time.

Referred to as the *head-related transfer function*, or HRTF, our head gets in the way of some sounds reaching the farther ear, and the pinnae reflect the intensity of the frequencies we hear. Perhaps head-related *transform* function would have made a more accurate name—since the term refers to how the sound is transformed by our unique bodies. We can localize high-frequency sounds more easily than low-frequency sound, because the high frequency sounds tend to get filtered out by the head. Low-frequency sounds are harder to localize because no acoustic shadow is created by the head. These spectral cues are also a key component in how we determine sound location.

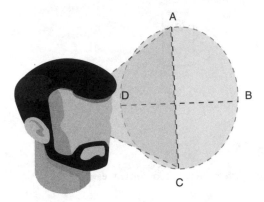

Figure 7.3
The cone of confusion.

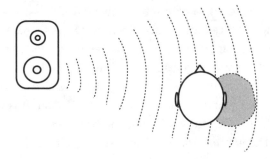

Figure 7.4
Spectral cues altered by a head shadow.

Other aspects of localizing a sound source also come into play, such as the initial time delay between a sound and early reflections, the amount of direct versus reverberant sound, and auditory motion parallax (when listeners move, sounds from sources close by appear to move more quickly than sounds from farther away—see Yost 2018), but these are less commonly used in artificially simulating spatialization.

Exercise 7.1   Where Does Sound Localization Fall Apart? (Partner Exercise)

An easy way to demonstrate where localization tends to weaken is to sit, blindfolded or with eyes closed, and have someone else snap their fingers in a variety of places around your head: is there a place where you can't figure out where the sound is coming from? Did your partner indicate where you went wrong?

Exercise 7.2   360 Degrees of Sound

Stand in one spot and point a directional microphone at the space. Slowly, over the course of several minutes, turn around on the spot, so you are recording 360 degrees of sound. Note your observations when you play back the file.

7.2   Binaural Audio

Binaural audio takes into account the main differences (i.e., in level, time, and spectral properties) between the perceptions of our two auditory inputs to our brain. These differences can't really be used in stereo or surround sound reproduction, because of what's called cross-talk: what is in each speaker will reach both of our ears. If we pan a sound hard left into the left speaker, our right ear will still hear the sound because it's traveling through the air and can't be isolated to one ear.

When we use headphones, we *can* isolate the sound and eliminate cross-talk. Binaural audio, then, relies on the listener using headphones, which is why it hasn't been used a lot in media, although you'll find at least one film (e.g., *Bad Boy Bubby*), a handful of music albums (e.g., Pearl Jam's *Binaural*), and an increasing number of video games (e.g., *Papa Sangre*) have made use of binaural audio.

Sound can be recorded binaurally using two microphones, but this requires the head to block some frequencies of the sound, provide distance between the two cues for

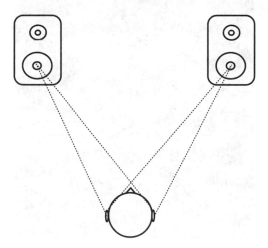

Figure 7.5
We hear each speaker with both ears, leading to cross-talk.

the level and time delay, and ideally pinnae that reflect sound. For that reason, when recording binaural audio, we can use dummy heads, which have been built to mimic a human head, or by using tiny microphones in our own ears (some people secure microphones to the side of a hat, but then they will miss the spectral changes created by the pinnae). More commonly, it's easier to create binaural audio in post-production by positioning the sound using binaural plugins for audio software. These plugins use a *generic HRTF*—an averaged-out sense of how people hear (perhaps most commonly, those datasets found on MIT Media Lab's KEMAR or IRCAM's Listen HRTF Database). That generic aspect means it's not quite as accurate as our own unique pair of ears, but it is still more accurate than stereo. I could find no native plugins for binaural audio for Audacity, but you can find some VST ("virtual sound technology") plugins that you can install into Audacity that will enable you to explore binaural audio. The microphone and headphone manufacturer Sennheiser have released a free binaural VST plugin called Ambeo Orbit, and although I couldn't get it to work in Audacity it worked fine in other software.

Exercise 7.3   Listen to a Binaural Recording

Check out some binaural mixes from music, film, or games. What do you notice about the difference in localization versus stereo?

Figure 7.6
Ambeo Orbit (running in Adobe Audition).

Exercise 7.4   Binaural Recording

If you don't have binaural mics, tape a mic to each side of your hat by your ears, and head out on a sound walk, recording as you go. Listen to the file when you get back. What do you notice is the difference between this recording and others you've taken on your sound walk? Did the binaural aspect work even without the pinnae being involved?

Exercise 7.5   Binaural Panning

If you've got a binaural plugin for your software, try to recreate a scene from your sound walks using binaural panning.

7.3   In-Head Localization

While it may be easy to think we can drag-and-drop our sounds into position for a stereo or surround mix, the potential use of headphones by our listener presents a challenge for most mixers because the lack of cross-talk leads to what's called *in-head localization*—the feeling that the sound is coming from inside our head. As we saw above, the front speakers in surround sound, or the two speakers in stereo sound, are optimally positioned to create an equilateral triangle, where the head is located at the sweet spot of an even mix from right and left channels. In factory mixer settings, for instance, the stereo imaging is usually set by default to this 60° angle.

Almost all headphones are designed to transmit one of the stereo channels to one ear exclusively (that is, the left channel transmits to the left earphone), so it is possible for one ear to hear an entirely different sound than the other ear. Sounds panned hard into one channel for speakers can therefore sound unnatural when translated to headphones, since in the natural world we do not normally hear a sound with just one ear. Some cross-feed software plugins simulate the loudspeaker experience with audio mixed for headphones, which bleed some of the left and right channels together. However, these plugins can still result in a not entirely natural feel.

When we eliminate cross-talk, the optimal positioning of sounds becomes much wider, and is typically tripled for headphone mixes (up to 180°). The reason for that wide spread is that in the equilateral triangle mix, the sounds appear to emanate from inside our heads, rather than from around us, and generally this is considered bad by mixers—we want to feel like we're standing on stage with the band, not like the band is playing inside our heads.

Exercise 7.6   In-Head Localization

If you have access to stereo panning tools, try to place sounds intentionally in and outside the in-head localization position. How does it change the way that you perceive sound? Reflect on where you might want to use in-head localization on purpose.

7.4   Surround Sound

Because most people don't use headphones to listen to film, most film is mixed for some form of surround sound. Surround sound can mean different speaker setups, and there are different standards based on what setup you are using. The basics of surround sound is to have more than the two standard stereo speakers, but how many additional speakers is fluctuating. In the 1970s the standard was quadrophonic, or four-speaker sound. Since the 1990s the standard for home theater has been 5.1, where "5" means 5 main speakers and the .1 means a subwoofer. Some new formats introduce height speakers as well, since in standard surround the speakers are all at the same azimuth. These can be called 9.1.2, for instance, where the ".2" at the end indicates the height speakers. You may come across 7.1, 9.1.4, 22.2, and more.

The basic setup is to keep the two front speakers (front left and front right, or FL and FR) where they are located in stereo position, each at 30° from the listener, making a total 60° equilateral triangle (figure 7.7). The center (C) is in the direct center, 0°, and the two rears, or surrounds (SL and SR) are each at 110° from the listener, to form a triangle at between 100° and 120°, or twice the width of the front speaker. This is sometimes referred to as the ITU standard (it's ITU-R BS.775, to get technical). It's not the only standard, however: the National Academy of Recording Arts and Sciences, NARAS, recommend the rear speakers be between 110° to 150°. For home theater, Dolby and THX recommend the rear speakers be between 90° and 110°. Because the subwoofer's frequencies aren't localized well, you can place your sub pretty much anywhere.

There are three major competing formats for surround sound: Dolby, DTS, and THX. Each of these has its own unique recommended placement of speakers, and if you've got a home stereo system with separate components (rather than an all-in-one or Bluetooth setup), you'll probably find options on your receiver to use Dolby, DTS, and/ or THX. Each compresses audio differently, and each uses different means to decide which channel to send information to and which to apply equalization to. You may come across Dolby Digital, Dolby TrueHD, Dolby Atmos, DTS:X, DTS HD, THX Ultra2, and more. You will need to look up the specifications for setting up your equipment.

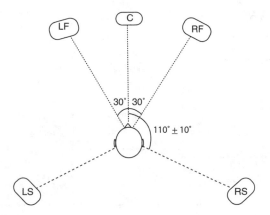

Figure 7.7
Standard surround sound setup.

The differences are a matter not just of speaker location, but also how they play back sound. We know that bass response drops off at lower amplitudes, for instance. In theaters, the volume is high so we have plenty of bass response. At home, you're probably not going to crank up your volume (at least, if you live in an apartment), so you are going to miss out on some of that bass. By EQing the mix in decoding, we can reintroduce the illusion of amplitude, as we have seen. In other words, how the setup uses EQ can alter your experience of the sound.

Exercise 7.7 Test Surround Modes

If you have the equipment, grab a great DVD or Blu-ray (yes, a real physical DVD, not a stream—ideally you can get your hands on a *demo disc* on eBay, designed to demonstrate the capabilities of the system) and try out different surround modes. Can you hear a difference? What are the most obvious differences you hear?

Exercise 7.8 Surround Mix in Audacity

It is possible to mix in surround in Audacity. You may need to enable the plugin. With three or more separate tracks, select `File > Export > Export as Wav` then hit save. The Advanced Mixing Options window should pop up. This lets you map the tracks to output channels (figure 7.8). You can click on the track name followed by the channel number to alter the track's mapping. Mix a bunch of files in surround sound: how does it change your listening experience?

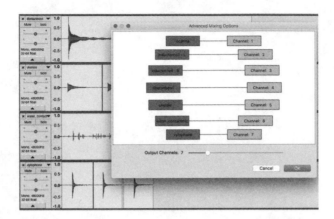

Figure 7.8
Advanced Mixing Options in Audacity.

7.4.1   The Exit-Sign Effect

Named by film sound designer Ben Burtt, the *exit-sign effect* is a phenomenon that relates to discrete sounds that are placed in rear surround loudspeakers. With film or other audiovisual media, pans of fast-moving sounds to the left or right of the screen boundaries lead the viewer's eye to follow the sound through the *acousmatic* (off-screen) space and toward the theater's exit signs at the side of the cinemas. For this reason, when we have a viewer sitting facing a screen, we don't tend to put spot sounds in the rear speakers (unless we want to shock an audience and have it appear that something is suddenly behind them). However, with just audio media, since it doesn't matter where the viewer is looking, we can use spot sounds to our heart's content.

7.5   Ambisonics and Object-Based Audio

Unlike surround-sound formats, which typically put speakers on one plane (one azimuth), ambisonics is a format that has been around for a long time that is designed for a full sphere. Rather than encoding information for each speaker channel, ambisonics creates a virtual sound field called a *B-format* that is decoded at the consumer end based on the individual speaker setup. The creator, then, doesn't worry about assigning sounds to specific speakers, but rather to a spherical position. The actual speaker location is chosen by the decoder and depends on the listener's system.

Dolby Atmos takes a similar approach, by allowing creators to position sounds in a 3D spherical position without attention to assigning speaker channels, and the sound is then decoded on the consumer side depending on the individual setup. Other similar formats have been emerging as spatial audio gains interest, including DTS:X and Auro-3D. These formats are often referred to as *object-based audio*, since they are independent of speaker channels, but position sound objects in a virtual space. It's a bit like visually placing an object in a room relative to us standing in the middle, and then placing that template into any room, so that the sound is always in the same relative position to the listener, no matter how large the speakers or where they are in the room. In this way, the creator need only position the sounds once and they are appropriately decoded on the user's device, whether they have a surround sound setup or are using headphones. The advantage for the creator, of course, is not having to create multiple deliverables.

New tools like object-based audio often mean new challenges. If you think about film ambiences, we used to create stereo ambiences and then just kind of leave them in the surround speakers. Now, we have to think about spatializing those sounds in three dimensions. A helicopter rising overhead could sound like it's actually flying right over our heads. But we still have to be aware of the exit-sign effect and not make spatialized spot sounds so obvious that they pull a person's attention from the screen. Of course, if we're not designing sound for film, then the challenges of object-based sounds don't become challenges so much as interesting creative opportunities.

To work with Atmos, DTS:X, or Auro 3D you'll need specialized tools, but Ambisonics is not proprietary, and there are many tools that you can find to explore Ambisonics. Most of these are open source or designed by researchers, so may not be as user-friendly as some of the proprietary tools. Two free options that are designed by major companies who do sound for VR/games are Facebook's Spatial Workstation and Audiokinetic's WWise. It is possible to encode an ambisonic format in Audacity using the Advanced Mixing Options.

7.6 Spatial Sound

There are many spatial sound ("3D") algorithms and approaches, and more advanced software is required to use spatial audio, so we won't go into too much depth here. Rather than referring to the position of the speakers or the channel assignment, *spatial audio* usually refers to the processing done on the sound source itself before it gets to the speakers/headphones, so it is usually used in virtual environments like VR and games. The goal of spatial audio is to create a three-dimensional position for the sounds. Note that with surround sound, the speakers are usually on one plane (that is, the same azimuth; although newer approaches to surround have a second height set added, as described above), and most sounds happen in the front speakers. With spatial audio, we have a 360° spherical positioning of sounds in the software. How that positioning gets decoded into surround setups or headphones depends on the decoder used, as we saw above.

7.6.1 Head-Tracked Spatial Sound

When watching movies, we are in what is called a *head-locked* position: our eyes are pointed forward and we aren't moving around much. However, in virtual reality (VR) applications, we can turn our head around, and our field of view changes with the position of our head. For this reason, the audio also needs to change. If there is a bird chirping on our left in VR, and we turn to look at it, the bird's chirp should now be in front of us. So, *head-tracked* audio that locates where our head is positioned can be a big element of realism in virtual reality spatial audio. Real-time, head-tracked spatial audio can be quite computationally expensive (that is, we have to have powerful processors), and so we often need to make compromises in the accuracy of the audio presentation in VR today, but this is quickly changing with GPU-based processing.

7.7 Sound Propagation

An important element of spatial audio rendering is understanding how sound waves propagate—how they travel through space, how they reflect off objects or are absorbed by objects, and how they decay over time. We looked at some basic propagation earlier (chapter 4), but with spatial audio we have to consider even more potential aspects of propagation. The wave's path through space, including all of its reflections and absorptions, must be simulated by the software for a realistic presentation.

Diffraction is the path of sound around obstacles. For instance, a sound that occurs behind a small opening can re-radiate as if it were a new sound source, depending on frequency.

With *occlusion*, the direct path of the sound is obstructed, and the reverberations/reflections are also muffled. Common occlusions include, for instance, walls that are between the sound source and the listener.

With *exclusion*, the direct path is clear, but reflections are obstructed in some way: a door in a wall, for instance, will let the direct sound through, but many reflections may be behind the wall.

With *obstructions*, the direct path is obstructed in some way, but the reflections can get around the obstructing object. Examples might be a large tree, a column, a car, and so on.

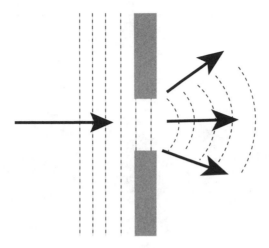

Figure 7.9
Sound diffraction through a small opening.

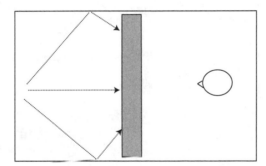

Figure 7.10
Occlusion effect.

Sound engines designed for spatialization must take into account these diffraction behaviors of sound as it propagates in the virtual space, if the goal is to create a space that is true to life.

Exercise 7.9   Real-Life Propagation

Use your home or school to set up some real-life propagation experiments. Play a broadband (many frequencies) sound from a speaker behind a door, then open the door. Duck behind some furniture. Note how the sound changes in the space.

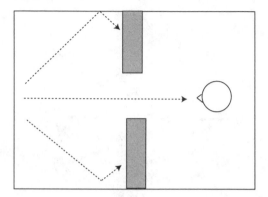

Figure 7.11
Exclusion effect.

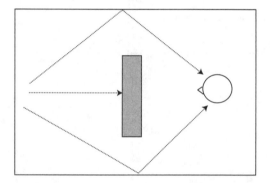

Figure 7.12
Obstruction effect.

Record the sound and run a spectral analyzer. What frequencies drop out as the location of the source changes?

Surround and spatial sound is not just a technical tool, but also a creative tool, as Stephan Schütze explains:

I actually think it's a lot more important than unfortunately many of us actually have the opportunity to make use of. And I'll give you a couple of examples. I've always played games, and I love games, and I'm probably addicted to them and that's why it's good that I work in the industry. And so in a lot of the studios that I worked in, we would play the games at lunch time or after work, etc. And because I had my own little space and because I was mixing these in surround I had very, very simple surround setup initially, but we would play the games that everybody else was playing, and at one point we were playing *Counter-Strike*. And there was a particular level where there was a particular little nook where you could hide, and I would hide there occasionally, and one day somebody very, very carefully crept up behind me, turned the corner, and I shot them. They were like, "How did you know I was there?" I said, "I heard you coming." He said, "But I was really, really quiet." "Yeah, but I've got rear speakers." And as a sound designer, I'm very, very aware that the sound of you walking through the corrugated iron tunnel to get to me has a very distinct sound. So as soon as I hear a corrugated iron tunnel anywhere in the level, I know there's somebody in the corridor behind me. But the sound, the surround sound pinpointed that exactly.

I do find it really interesting that there are people playing really competitive, high-end competitive first person shooter games with stereo headphones, and I'm like, "What? Your critical sense for threat analysis and warnings is your hearing." They say your ears don't blink. They don't go to sleep, either, really. And so surround sound, from a tactical point of view, is brilliant, but from an environmental point of view, if we want to simulate an environment, you've gone out to a beautiful jungle or a beautiful forest somewhere, or even a desert, with insects, and it's like you've got your head underwater. In the same way that water completely envelops you, sound completely envelops you and you listen to a really amazing movie in surround sound. You take those speakers away and all of a sudden it's like looking at a flat painting. If you take so much out of the experience and for games, it's experiential. It's narrative support. It's tactical opportunities, there's so many things in there, and from the point of view of creating an audio environment. It's like if I was an artist with, somebody sort of said, "OK, here's your pencil. Draw this beautiful picture." And I'm like, "OK, cool, I'm ready to colour it in, can I have the colour pencils?" "Like, no." I was feeling like I'm missing all the colour when I don't have the ability to do things in surround. (quoted in Collins 2016, 198)

Reading and Listening Guide

Benjamin Wright, "Atmos Now: Dolby Laboratories, Mixing Ideology and Hollywood Sound Production" (2015)

Wright's article is an academic study of some history of Dolby Atmos and surround sound, and then some approaches used by Hollywood film mixers to surround sound and Atmos in particular.

Although quite technical, some of the best information about sound spatialization techniques can be found in the documentation available for software tools such as the game audio middleware tools Wwise and Fmod.

You'll find binaural audio in an increasing number of video games; for instance, *Papa Sangre* (2010), *Hellblade* (2017), and *Sniper Elite 4* (2017). Pearl Jam released an album recorded in binaural audio (*Binaural*, 2000), as did Can (*Flow Motion*, 1976). You can also find some classical music recorded in binaural sound. If you have the playback technology, you can also switch between listening to media in standard stereo and in surround sound, so you can get a sense of what the surround speakers bring to an experience. Some music is also mixed in surround formats. For a while in the 1970s, quadrophonic (four-speaker, or 4.0 surround) music was popular with many artists (you'll have to find a copy of the quadrophonic version, not the stereo version, to listen to it). In the late 1990s and early 2000s, 5.1 surround music albums were also released for a short time, as "super audio CDs," or SACDs. Now that most people have shifted to streamed music, you can pick up some SACDs cheaply, along with the used hardware required to play it.

8   Sound and Meaning

Every single sound, no matter how subtle or simple . . . is considered, composed [and] orchestrated to have an effect on the audience.
—Mark Mangini, sound designer (quoted in Horn 2016)

What and how do sounds mean? We've now got a grasp on what sound is, how to use sound libraries, and how to record, edit, and mix sounds; but how do we know what sounds to use, and where? Understanding how sound creates meaning will help you to understand how best to use sound to creative advantage in your projects. A very creative use of sound can also sometimes make up for less technical proficiency.

There are many theories about sound and its meanings that come from different disciplinary perspectives (psychology, philosophy, musicology, and linguistics, for instance). It's not my intent to cover all of the theories of sound and meaning, but understanding some of the major theories about sound and meaning will help you to understand, talk about, and use sound more effectively and more creatively. At the heart of each of these theories is the concept that sounds are complex and evocative: that is, they are never just "sounds," but become associated with other things through our daily experiences of them, and these associations are critical to our understanding of how to use sound as a component of design.

8.1   Conditioning

Conditioning is a psychological behavioral process in which our response becomes learned over time as a result of being reinforced by repetition or extreme circumstances. The classic example is the Russian physiologist Ivan Pavlov, who in the 1890s ran studies of his dogs responding to being fed. In these studies a bell was rung every time the dogs were fed, and the dogs came to associate the sound of the bell with food. Pavlov could then ring a bell and the dogs would salivate. Today, we continue to use conditioning in training dogs for bomb and drug sniffing, but humans are *also* trained to respond to certain events in certain ways, consciously or not.

Conditioning is used in product, film, and game sound to make the listener associate sound with certain events: watch the movie *Jaws* (1975), and notice how we're conditioned early on that the famous musical cue represents the shark. When we become conditioned to expect the association, that link is broken on purpose, and the shark comes without warning, leaving us off-balance and adding tension. In video games, we're often taught to associate certain sounds with certain events—that was a bad poison we just drank, or it's good to collect coins. The association of sounds with events helps teach us or orientate us in the game, reducing the learning curve. And we're conditioned in other aspects of our lives by sounds—have you ever been out somewhere and heard someone else's phone with your ringtone? Did you jump or grab your own phone?

With a number of colleagues at the University of Waterloo I've studied the use of sound in slot machines to manipulate and control players. In slot machines, of course, we only ever hear winning sounds: we don't hear losing sounds. But we discovered that slot machines also play winning sounds when we lose sometimes, what we called "losses disguised as wins" (see Dixon et al. 2014). A loss disguised as a win is when, for example, we put in a $2 bet and win back a dollar. Technically, we lost a dollar, but the sound tells us we won and plays the same winning sounds as a real win. We respond, physiologically, to these losses as if they were wins, because the winning sound conditions us to hear them as wins. In this way, sounds are deliberately manipulative as they play on that conditioning effect.

It doesn't need to be exactly the same sound each time to evoke a conditioned response. Similar sounds can affect our perception of a sound: one scientific study conditioned mice to associate closely related sounds with trauma (see Aizenberg and Geffen 2013). By playing a similar sound, they could invoke a similar response in mice.

The theory of classical conditioning—the Pavlovian notion that we could be conditioned to respond in a particular way—gave way to a more complex understanding of this process under B. F. Skinner, who introduced the idea of *operant conditioning*. With operant conditioning, there are positive and negative reinforcements, and positive and negative punishments. "Positive" and "negative" refer to the presence or absence of something, rather than good or bad. For instance, a positive reinforcement is a reward—like a sonic reward in a game. A negative reinforcement is the removal of something—like the seatbelt sound your car makes until you do the desired behavior (plug your seatbelt in). When it comes to punishment, a positive punishment is adding a negative consequence to decrease a behavior—like an angry buzzer sound in a game show when you do something wrong. A negative punishment is to take away something good to reduce a behavior, like revoking your access to music that you love until you change your behavior.

Exercise 8.1   Conditioned to Sound

How many sounds can you list that you are conditioned to respond to in a particular way? What is your association and why? What types of conditioning are they (positive/negative, reinforcement/punishment)?

Exercise 8.2   Conditioning Sounds

Try to imagine some creative uses of the conditioning effect beyond some we've already talked about (hint: think about products, advertising, etc.).

8.2   Sonic Archetypes, Stereotypes, and Generalizations

Archetypes are a kind of ideal example of a concept, and show up in dreams, literature, religion, and art across the world: psychoanalyst Carl Jung, with whom archetypes are often associated, called them *primordial images*. For Jung, these archetypes show up as personality types or patterns of behavior, like the hero, or the wise old sage. These differentiations between types of people are often used in writing for fiction. They each have a defined main set of desires, fears, and talents. Because they are so often used in storytelling, it's useful to understand them and think about how we might represent these people in sound.

Here I use *archetype* in the broader sense of the term, as patterns or prototype examples that can represent a phenomenon. Some of our archetypes have developed through evolution and biology, and others are cultural. Some sounds carry *universal* meaning to all humans everywhere (*global* would perhaps be a better term). Low-frequency sounds are typically associated with threat, because those are the sounds that signify danger in nature: thunder, volcanoes, earthquakes, large mammals. High-frequency sounds are not usually associated with threat: small birds and the like. There are exceptions, of course. When high-pitched sounds are associated with threat, they are typically of the psychological disturbance variety, such as Bernard Herrmann's strings in *Psycho* (1960) (see Collins and Tagg 2001). These sounds have become archetypal in film: it's very rare to have a scary scene without a low-frequency hum or a screeching high frequency accompanying it.

We also have *conventional* sounds: sounds whose meaning we agree on through cultural convention, not through some universality. For instance, the sound of a phone ringing: there is nothing inherent in the sound to indicate that a phone is ringing, but everyone in the West understands what a ringing phone sounds like (a doorbell ringing is another example). Our conventions tell us what the sound means—what it refers to, or what it *connotes*. Conventional sounds can be archetypes as well, although their meaning is specific to specific audiences. The sine waves accompanying early computers as warning beeps became standard interface sounds in science fiction, and remain so today, even though we could use any sound now.

Beyond these universal and conventional sounds that are biologically or culturally reinforced, we also have sounds that are personal to us based on our past associations with them, as we saw with conditioning. On a personal level we associate sounds with certain events in our own lives. We can use the term *anamnesis* to describe this phenomenon, the "involuntary revival of memory caused by listening and the evocative power of sounds" (Augoyard and Torgue 2009, 21). These may be shared by groups of people who experience similar events, but may not be a widespread cultural phenomenon. There are times I've heard a sound similar to the sound Super Mario makes when he jumps and gets a coin, and I immediately am reminded of playing the game with my younger brother.

If we are designing sound, it's important to know our main audience. Composers and sound designers often use sonic symbols that signify other cultures, but those cultures can be offended by such generalizations or have different associations with those sounds: perhaps every time we go to "Australia" in a sonic narrative, we hear the didgeridoo, for instance, but Australians may associate the didgeridoo not with Australia, but rather with the indigenous Northern Australian population specifically. It's important, then, to understand who our audience is and what associations they may have with the sounds. We can't control the anamnesis, or personal responses to sounds, but we can control which sounds we choose, based on how people will respond to or understand conventional or universal sounds.

Exercise 8.3   Universal and Conventional Sounds

Listen to different aspects of a soundscape you've recorded. What aspects are conventional and which are universal? Download thirty random sound effects and categorize them as universal or conventional. Explain your category choice.

Are there any features of the sounds that are unique to conventional or universal sounds?

Exercise 8.4   Sonic Generalizatons and Stereotypes

What sonic generalizations can you recall that represent places, cultures, religions, seasons, periods of time, or other aspects in film, television, radio, or games? Think beyond just culture: for instance, every time we see a bicycle on screen in a rural setting we tend to hear a bicycle bell. Nighttime often has the sound of crickets, or if it's urban, a police siren. What sounds are associated with women? What sounds with men? How do the sounds of women differ from those of men, and why?

Exercise 8.5   Anamnesis

Are there sounds that you know are not conventional or universal, but which affect you on a personal level by evoking some past memory? I associate a mourning dove with my grandmother's house, for instance, since when I stayed at her house I would often hear them. Make a list of all the sounds you can think of that cause anamnesis for you, along with their associations. Keep adding to the list as you think of more over the coming days and weeks.

Exercise 8.6   People Sounds

Sit in a public place and watch people in the vicinity, and mentally assign them their own sound effects. Alternatively, choose people from your own life and think of them in terms of their own sound effects—imagine assigning them a sound effect as a ringtone identifier. What did you choose and why?

8.3   Basic Semiotic Theory

Semiotics (sometimes called *semiology*) is the study of signs and symbols, and has developed from linguistic theory as an understanding of communication and language. It can get quite complicated, but it's a useful way to understand sound as a form of symbolic, rhetorical, communicative language. The field uses different terminology depending on which theorist we're looking at, which can make semiotics confusing. We're going to stick to a very simple explanation here, but semiotics can be a really fruitful theory to use to talk about and analyze the use of sounds in media, and if you're interested in how sound communicates meaning on a theoretical level, I'd recommend you dig a little deeper into this subject.

In the semiotics of Swiss linguist Ferdinand de Saussure, anything that represents something else can be referred to as a *sign* or *signifier*. What that sign refers to—the idea, concept, or meaning—is known as the *signified*. If we take the sign of the sound of a phone ringing, the ringing sound is the signifier, and the signified is that someone is calling. We could take this further: if, for instance, I am waiting for a dreaded call from a doctor to find out if that lump is malignant or not, the signified might also include anxiety or other emotions. If we imagine a film scene in a thriller where something bad is about to happen and a car alarm goes off, the car alarm might signify that someone is breaking into the car (or just bumped it), but it can also be a signifier of alarm and distress for the character in the scene about to get attacked.

The terms *denotation* and *connotation* are used to describe the relationship between signifier and signified. In the above example with a phone ringing, the denotation was "someone is calling," and the connotations were anxiety and dread. The denotation is the literal meaning of the signifier, and connotations are the secondary associations we have with the sign. In everyday use, we don't often speak of denotation, because that is usually quite obvious, but the connotations are what are important in sound design. What a sound means—beyond its literal representation—can add so much to a scene. Imagine a film scene in which a character is getting increasingly angry while having an argument in their kitchen. The simultaneous use of a kettle boiling and coming to a scream might *connote* the feelings of the character.

The fact that sounds may carry information means that they may also cause a communication failure of interpretation. We may not have the same associations and connotations as our audience, so it's important to understand that whatever signs we use, the audience understands the meaning. As Umberto Eco puts the matter: "To make his text communicative, the author has to assume that the ensemble of codes he relies upon is the same as that shared by his possible reader. The author has thus to foresee a model of the possible reader . . . supposedly able to deal interpretatively with the expressions in the same way as the author deals generatively with them" (1979, 7). Putting that into our own terms, the listener first has to understand that we're using sound as a metaphor or symbol, and second has to understand the sound with the same interpretation as we would (on a conscious or subconscious level). For instance, they would have to understand not only what a kettle boiling sounds like, but also the association of boiling over with stress or anger.

As well as sharing the same symbols, the listener and sound designer must share the same expectations and sociocultural norms. In other words, the listener and sound designer must agree that the symbols represent the same sociocultural ideas, and the listener must have an adequate response to those ideas. For example, if my father heard the heavy thrum of a rap song's beat, he might associate this concept with the teenage Jeep jockeys with massive woofers who drive by his house late on a Friday night and wake him up. Although he therefore may have the understanding to hear the sound as representative of some form of power, my father will likely hear this as somebody imposing their power on others, while the teens may hear this as making them feel powerful.

Ultimately, when we use sound, we should think about its potential meanings for different audiences. If we have a personal association with a sound, we can't assume everyone has that same association; we have to consider other potential meanings. Most of the time, sound designers do this intuitively, but consciously thinking about sound's potential meanings or connotations can be a fruitful way to explore the use of sound as a metaphor.

Exercise 8.7   A Sonic Warning

A classic visual design exercise is to draw a sign that anyone in the far future could understand to mean that a place is very dangerous, like a nuclear waste disposal site. The design problem is what images one could use to signify the danger of the site in order to be universally understood. How would you use *sound* to represent a universal warning for the future? Make that sound. What effects did you rely on and why?

8.3.1   Semiotic Analysis and Reference Material

While we may not consciously use semiotics in our sound design practice, we may find ourselves using some of its techniques without realizing it, particularly when trying to think through why a particular sound or particular group of sounds evokes in us a particular response. We can analyze what sounds mean by comparing uses of that sound to similar scenes or similar sounds in other media. Musicologist Philip Tagg (2003) calls this *interobjective comparison* (that is, comparing between objects, or texts, like books, film, and radio). For instance, if we find the same doomy bass sound right before a murder in six different movies, we could probably surmise that the doomy bass signifies murder, threat, and anxiety. We might also gather similar sounds from sound libraries and compare the terminology used to describe the sound: if the sound effects were given names like "murderous bass" or "threat," these are in the same general connotative sphere, so we could gather multiple examples of how people associate certain sounds with certain responses. If we want to signify an alien landscape, the first place we might look are sound libraries with tags like "alien." We might go watch and listen to sci-fi radio plays, movies, or television. That way, we have an idea of how other people have already created a template of signification for us to follow.

Another approach involves a comparison not between objects or uses of sounds but across audience responses. These are inter*subjective* comparisons, and take a bit more work, because they require tests on a real audience. Years ago, I explored sound connotations by using *reception tests*. These are kind of like an auditory Rorschach ink-blot test: you play a sound, and then ask people to respond freely to what the sound connotes for them. In my case, I played them common sounds used in science-fiction cinema, and asked them to give the sound a name, describe what made the sound, and any associations they had with the sound (Collins 2002). Many responses described the physical causal relationship they heard in the sound: "stampede on sheet metal," "metal and ironmongers," "arrow shot from laser gun." Such images indicate that, far from sounding physically detached, sound samples are "seen" in the mind as having activity and motion (we'll come back to that below). In fact, remarkable in my study that a single sound would call up in subjects' imagination a specific film. This told me that single sounds can carry a lot of weight when it comes to signification. The listeners weren't aware of where the sounds came from or what I was looking for, but many found connotations very close to the meanings I had perceived myself, and confirmed with the interobjective methods I'd used as well.

Trevor Cox, a researcher at Salford University in the UK, ran a study online a number of years ago called *BadVibes* that asked people to vote on the most horrible sounds they could imagine. By collecting a lot of data about how people respond to certain sounds, Cox could determine what sounds people associated with particularly unpleasant feelings (the sound of vomiting came out ahead overall). Cox is often running more experiments on his website, sound101.org, and it's always worth spending a few minutes to help out researchers, because that information eventually helps feed back into training new practitioners. It also helps you to see how our understanding of sound as a communicative tool is developing.

Exercise 8.8   A Semiotic Study of Sounds

Find a sound composition: it could be a radio play, or the sound to a movie (without using the images), or a *musique concrète* composition. What are the important signifiers in the piece, and what do you think they signify? How do they relate to other signifiers used? Can you find similar sounds used somewhere else that support your idea of what the sounds connote (interobjective comparison material)? What stereotypes do you hear, and which are cultural or universal? Carry out a reception test among your friends, colleagues, or others and have them freely respond to the sounds. Did they all have the same connotations? What were the differences, and why do you think those differences existed? What happens to the sound's meaning when you put different effects on the sound?

Exercise 8.9   A Semiotic Study of a Soundscape

Create a soundscape, for example, the basement of a serial killer's home. Repeat your semiotic study as described above. What responses did you get, and how did you use sound to help evoke those responses? What effects did you add to any recorded sounds, and what impact did those effects have?

Exercise 8.10   What Do I Mean?

Create an alien sound language and devise a short message. See if anyone can interpret it. What research reference material did you use and why? How did you test its meanings?

Exercise 8.11   *Musique Concrète*

Musique concrète was a style of music that used recorded sound effects as raw material, back in the days of tape, and is exemplified by the works of Pierre Schaeffer and Pierre Henri. Listen to a piece of *musique concrète* and freely associate to it by writing down any imagery that pops into your mind as you listen. Listen to the piece again—what sounds or effects led to that imagery? Create your own piece of *musique concrète* that relies on sound objects that you think have specific connotations, then try it out on an audience and ask them to freely associate to the piece. Were you able to evoke the imagery you wanted? Why or why not?

8.4   Phenomenology, Embodied Cognition, and Intersensory Integration

One other major group of theories that is useful for our thinking about sound and its meanings relates to areas of philosophy and psychology that have tried to understand the senses and the body in relation to the world and its meanings: phenomenology, embodied cognition, and intersensory integration. These are all related, as they all deal with our senses and our understanding of the world through those senses.

Phenomenology is a branch of philosophy relating to how human experience is mediated by our sensory perceptions and our bodies. Phenomenologists hold that our mind and how we think about the world is shaped by our physical experience and existence in that world. German philosopher Martin Heidegger (1962), for example, believed that our experiences are predicated on our being embedded in our environ- ment (what he calls *being-in-the-world*). In this way, how we think is shaped by our interactions with the world. French phenomenologist Maurice Merleau-Ponty (1998) similarly held that all perception is understood through the ways in which we are able to act in the world and move around in our environment. In this way, the sensory inputs that we receive from the environment mold our understanding of the world. Our body acts as a mediator between the world and our consciousness.

Phenomenology has been used in music and sound studies to explore the ability of sounds to evoke imagery in our minds, as a means of explaining some of the semiotic connotations discussed above. For example, in exploring radio drama, Clive Cazeaux argues with the German film critic Rudolf Arnheim, who stated in the 1930s that radio "seems much more sensorily defective and incomplete than the other arts—because it excludes the most important sense, that of sight" (Cazeaux 2005, 158). Cazeaux used the phenomenology of Merleau-Ponty to argue that our five senses are not five discrete channels that gather data for the mind but, rather, that the senses operate in unity: we listen and simultaneously see and feel the drama. And although we always listen causally, we are simultaneously interpreting and associating sounds with other events, objects, and emotions from our experiences.

Some phenomenological approaches to music and sound suggest that sound alone can evoke bodily sensations of touch and movement: these views include the "mimetic hypothesis" (Cox 2001), "bodily hearing" (Mead 2003), and "vicarious per- formance" (Cone 1968), among others. If I asked you to perform out loud the action and sound from the infamous *Psycho* shower scene, you'd probably lift your arm and say "eee—eee—eee." I've seen people use that motion and sound as shorthand for say- ing someone is a little unstable. But if you watch the scene, you'll see that actually the stabbing is not timed at all with the screeching strings. So why do we remember that sound as being linked with that action? The answer has to do with kinesthetic sympa- thy: even if we don't play the violin, we have an understanding of how the violin is played, and we hear the action of the players in the music—a very stabby action! When we remember that scene, with the action of seeing the stab, even though it's not timed to the strings, we remember them as being tied because of our bodily hearing of the music cue.

We understand human-made sounds (including those of playing a musical instrument) in terms of our own experience of making similar sounds and movements. That mental re-creation of the sound causes a neuronal and sensorimotor response that mimics the performer's actions, and then we are able to interpret the emotional inflections through a mental re-creation of that action. People therefore can give meaning to sound in terms of mentally emulated actions. Put differently, we mentally (and sometimes even physically) imitate the expressiveness of an action or feeling behind a sound, based on our prior embodied experiences of sound making. I mentioned this above in my own studies of sound effects—some people freely associated actions behind sounds in their descriptions of imagery evoked by sounds.

Embodied cognition is the cognitive science corollary to phenomenology, and is based on recent cognitive science being done with fMRI machines. Scientists are studying how our visual, auditory, and physical systems in our brains are all closely linked, and the result is that we understand sound through our bodies as much as our minds. In the past fifteen years or so, scientists have discovered that when we've experienced something, that experience fires off neurons in our brain, so if we encounter a grizzly bear, a bunch of neurons are going to fire in our brain. If, later on, we just see a movie of a bear, or hear that roar, almost all the same neurons fire. In other words, when we get a similar stimulus, our brain pretty much responds as if it's happening to us, for real, again. This works so well that if we see something happening to somebody else, we still experience it in our brain to some extent as if it's happening to us. So if we see someone else getting pricked by a needle, many of our neurons fire as if we are being pricked by a needle, which is why they're called mirror neurons—we mirror the experiences of others (see Niedenthal 2007).

Phenomenology and embodied cognition provide support for theories of intersensory integration—the theory that our senses are not separate unique inputs, but have some overlap, as Cazeaux argued above about radio. Some people have quite literal sensory integration in that they have a neurological abnormality called synesthesia, in which they may "see" a sound or "hear" a color, for instance. Chromesthesia is a sound-to-color synesthesia in which heard sounds evoke an experience of color. While there are many theories as to why this unusual sensory experience occurs, we probably all experience the phenomenon to some degree. That is, we all have some intersensory integration—our senses are not discrete individual "inputs" to our brain, but are overlapped and integrated in our minds. Recent research has shown, for instance, that when your eyes move, your eardrums move too, bringing together sound and vision (Gruters et al. 2018).

Our senses are constantly exposed to stimuli from multiple sensory modalities (visual, auditory, vestibular, olfactory, taste, touch), and we are able to integrate and process this information to acquire knowledge of objects in our environment. The senses interact with one another and alter each other's processing and ultimately our overall perception. We're not going to focus on the actual interaction of sound and image in this book, but it's important to understand that image changes the perception of sound (and vice versa); so we can't just take an audio file we've created for a radio play and put it to images and expect the same responses from an audience.

The idea of intersensory integration with sounds means that the way our brains process sound relates to how we process our other senses at the same time in association with sounds, or because of our own bodily response to sound. For example, if you imagine lemons making sound, would lemons make a low-frequency, mid-frequency, or high-frequency sound? Most people would say that lemons make a high-frequency sound. But lemons make no sound, so how can we agree on what sound they might make? Now imagine yourself making that sound: close your eyes and sonify lemons. Our shoulders go up, eyes get pinched, lips draw back—we feel the sound in our body—the piercing high frequency is similar to what our body does when we taste very sour food like lemons. This demonstrates how sound and taste can be integrated in our minds, and those types of interactions occur between other senses as well. Marketing scientists have studied the association between sound and taste for years—"gotta have that crunch," for instance: the sound of potato chips affects our sense of their taste (see Spence and Shanker 2010).

Researchers have done a lot of work on the interaction between sound and image, particularly from the fields of psychology and marketing, and also from film studies. For instance, the *ventriloquist effect* refers to the idea that we localize sound with an image, particularly if that image is moving, as when we see a ventriloquist dummy "speak" even if the voice is coming from elsewhere (see Choe et al. 1975). Another well-known example is the *McGurk effect*, in which visuals can alter our perception of speech sound. The McGurk effect occurs when the audio component of one sound is paired with the visual component of another sound, leading to the perception of a third sound (McGurk and MacDonald 1976). Perhaps most well-known is the *Bouba-Kiki effect*, a phenomenon of *phonosemantics*, in which certain visual shapes are associated with certain consonant and vowel sounds: "bouba" is generally drawn as rounded, and "kiki" as having hard angles. These associations have been demonstrated across cultures (Köhler 1947). An interesting recent study has demonstrated that the Bouba-Kiki effect also occurs when the visual domain is occluded and the participant is given shapes to touch, rather than see—in other words, sound-shape associations hold for the haptic-auditory associations as well as visual (Fryer et al. 2014).

While visuals most often influence our perception of audio, audio also affects our visual perception. For example, beeping sounds can create the illusion of visual flashes (Shams, Kamitani, and Shimojo 2000). In another test, known as the *motion-bounce illusion*, two identical visual targets will be perceived as crossing through each other in the absence of sound cues, but when a brief sound is added at the moment that the targets interact, a bias perception toward the targets bouncing off each other occurs (Sekuler, Sekuler, and Lau 1997). Hulusić et al. (2010) showed that sound effects allowed slow animations to be perceived as smoother than fast animations, and that the addition of footstep sound effects to visual-based walking animations increased the perception of animation smoothness in a virtual environment. Such use of cross-modal illusions are useful when we need to drop frame rates or reduce visual fidelity.

In addition to evoking images, sound can also evoke touch, or haptic perception. Auditory cues frequently occur when we touch or interact with objects, and these sounds often convey potentially useful information regarding the nature of the objects with which we are interacting (Gaver 1993). Research has shown that when these auditory cues are presented without the touch, they can provide sufficient information for people to assess the size of objects and even what material they are made of (Freed 1990; Wildes and Richards 1988). In fact, research has demonstrated the ability of auditory feedback to effectively convey information regarding various attributes of objects including material, shape, size, and the like (Gaver 1993). Some studies have shown that we are able to discriminate touch-produced sounds from different surfaces when the sound is presented in isolation of other feedback (Lederman 1979). The majority of research to date indicates that people can discriminate between different objects and surfaces on the basis of their sounds (Lederman 1979).

Since sound is a physical vibration, we can feel some frequencies as much as hear them: most commonly the low-frequency sounds have a very physical feel, and if we put our hands in front of a loudspeaker and crank up the bass, we can feel the air move.

The implications of embodied cognition and intersensory integration are shown that sounds are capable of evoking responses in our bodies and minds based on our past experiences, and as sound designers we can draw on that knowledge to guide a person toward a particular feeling or association. Nick Wiswell, sound designer for many racing games, explains:

> If you're racing a car in real life, an awful lot of the driver's feel, and I've spoken to a lot of the race drivers about this, is not from the audio or it's not from the steering wheel, in some cases. It's the g-forces. You actually, you can feel the vibration of the wheels through the steering

wheel. And you can feel how the car is reacting to the g-forces basically from the seat of your pants. You sort of get that natural vibration and sense of movement, all of which is completely lost in a racing game. We can use force-feedback steering wheels to simulate the tires, but a lot of the cues a driver would use to let them know that things are going to happen they don't get. Audio is a big part of simulating that. When do you need to shift? Well, you can use audio. If you listen to the car's engine, you can pick out when I hear this tone, which it reaches this pitch, that's my time to shift. If you're listening to the tire audio, we've got lots of cues in there that are like, "You hear this sound? You're about to lock your brakes. You hear this sound? You're about to start losing traction on the front tires." This is if you're in a rear-wheel drive car, it's this sound and it's in the rear tires.

There's a lot of really interesting audio cues that we've put in there that are haptics-type feedback, but it's an audio feedback cue that's telling the player you're about to do this thing, stop doing that. Or, it's time to do something, so there's a lot of information in there that gives you that sense, and we do have a lot of low-end content in our recordings that if you've got a good subwoofer can simulate some of that feel. But people say, "Ah, a race car went by me in a race, and it really hit me in the chest." It's like, "Yes, it's 135 decibels. Of course it hit you in the chest." I don't get that when I'm playing it at home on my twenty-seven inch TV with little half-inch speakers. It's like, no, the laws of physics say that could never happen. So how can we make it sound that impactful, or that big, without it actually being that loud. I would like to think that if you did actually crank out a game to be that loud, it would hit you in the chest, but it's a physical response, and the subwoofer can play a part in that, and we can simulate a sort of pushing of the air. But unless it's at those volume levels it's never going to hit you in the chest like it does in real life. But there's a lot of work we can do there on simulating loudness. . . . Over the years we've used various distortion and saturation effects to try and make the car feel as loud as it is, even when you're not playing it at loud volumes. YouTube videos are mostly distorted, of cars, and people think that's how they sound, so you've almost got to slightly push things in that direction to make people think that, "Oh, it's so loud it's breaking up." We don't want to distort it and clip it and make it sound bad, but there are things we can imply to really sell that sense of volume and loudness. (quoted in Collins 2016, 312)

8.5   Summary

Whether we talk about semiotics, embodied cognition, phenomenology, conditioning, or other approaches to understanding sound, they all have at their heart that sound doesn't exist in isolation—it has associations beyond itself that we as sound designers can rely on. Analyzing our own responses to sound can be a powerful way to understand how we can use sound to get a desired response from our audience. While most sound designers do this intuitively, having a language to conceptualize and discuss this process with our design team can be a really fruitful way to discuss your work and why a particular sound works in the context of your product.

Exercise 8.12   Haptic Audio

Sit in a small space with a subwoofer and EQ your music to just play a series of low frequencies (<200Hz): can you feel the sound as well as hear it? Where in your body do you feel it? At what frequency does the sound become less physical and more cerebral?

Exercise 8.13   Sound as Shape and Color

What sounds match the shapes in figure 8.1 (adapted from Schafer 1992)? What colors should the shapes be? What does a harsh sound look like, compared to a soft sound? What is a cold sound? What is a warm sound? What is a yellow sound? A blue sound? What is a sour sound? A sharp sound? A round sound? Create or record the sounds and use any effects necessary to emphasize the correlation.

Exercise 8.14   Categorizing Sounds in Themes

Download thirty random sounds from freesound.org or from a sound library. Categorize them together according to broader themes (powerful/weak, bright/dark, good/evil, death/life, etc.): you can think about sounds according to colors, to shape (envelope), or connotation, for instance.

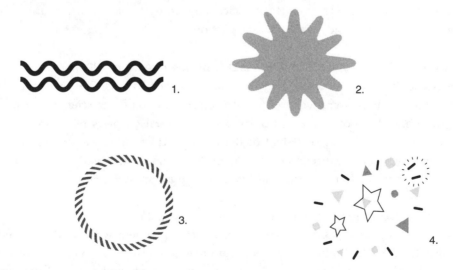

Figure 8.1
What sounds match these shapes?

Exercise 8.15   Changing Meaning

Take a sound and use effects to change its meaning. Make it dirty, wet, sharp, muddy, sad, evil, spikey, round, etc. Take one sound from the previous exercise, and using effects, transform it into its opposite.

Exercise 8.16   Classroom/Partner Game Sound Charades

Now that we have developed a language for talking about sound, we can play sound charades: Take a sound from your list of sounds and describe it in any way other than using a word related to what causes it. So a bird chirping could be described as a bright, cheerful sound, or as a sound that signifies dawn or spring, or in terms of its acoustic properties, like a high-frequency sine wave sound in short staccato bursts of about 60 dB. Have other students try to guess what sound you're describing.

Reading and Listening Guide

Karen Collins and Philip Tagg, "The Sonic Aesthetics of the Industrial: Re-Constructing Yesterday's Soundscape for Today's Alienation and Tomorrow's Dystopia" (2001)
In this article, Philip Tagg and I explore the semiotics of sounds in industrial music, a genre that relies on a lot of sonic conventions of science fiction cinema. One of the things we explore is which sounds are associated with dystopia and which are associated with utopia in soundscapes of science fiction.

Marc Leman, *Embodied Music Cognition and Mediation Technology* (2007)
Leman's book explores an embodied cognition approach to music research, bringing together research in cognitive science with research in computer science and musicology. While the book is focused on music and several chapters relate specifically to music technologies, the first half of the book provides an accessible overview of embodied cognition more generally. Leman outlines some of the issues with the more cognitivist approach to music that has dominated music study for centuries.

Trevor Cox, "Scraping Sounds and Disgusting Noises" (2009)
This article outlines Cox's psychoacoustic internet experiment with horrible sounds, and ties the sound to our biological and anthropological history. It's a fascinating read into why some sounds, like scraping nails on a chalkboard, evoke visceral responses

from us. Cox also explores how visual stimuli might affect our response to those sounds, in "The Effect of Visual Stimuli on the Horribleness of Awful Sounds" in the same journal. What are your most and least liked sounds, and why? Undertake any experiments Cox has going on at his website: why do you think he's testing those sounds, and what is the purpose of the experiments?

Philip Tagg, *Music's Meanings: A Modern Musicology For Non-Musos* (2013)
Tagg's thick opus is an overview of semiotics and musicology for non-musically trained people (non-musos). Although there isn't a lot of work on sound effects, there is a thorough overview of Tagg's approach to music semiotics. He provides whole chapters on intersubjectivity and inter-objectivity, on analysis. Regularly updated in the e-book version.

Don Ihde, *Listening and Voice: A Phenomenology of Sound* (1976)
Ihde is one of the leading figures in phenomenology, and though his writing in the 1970s is perhaps a bit too academic for today's tastes, his work is an important contribution to understanding sound from a phenomenological perspective. In this book he grapples with the dominance of vision in our language and our approaches to phenomenology up to that point in time.

9 Sound for Story

Fictional podcasts, audiobooks, audio dramas, and radio dramas are some of the most common uses of sound design for story outside of audiovisual media. *Podcasts* are episodic series of fiction or nonfiction news, tutorials, interviews, and stories, named after their popular consumption on portable MP3 players like iPods (today they are perhaps more commonly played on smartphones). *Radio dramas* are a more traditional format of fiction in single or episodic form broadcast on the radio, originally, but more recently through other media as well: sometimes these are also called audio stories or audio dramas, and include stories on smart speakers like Alexa or the Amazon Echo. Audio-only fiction, including audiobooks, increasingly don't just have an actor reading the book but include sound design as well. Even nonfiction podcasts take some form of storytelling, so I'll refer to these collectively as *audio stories*. Today, many audio stories are witnessing a wave of interest, as portability (MP3 players, phones) and easy access have made listening to podcasts and audio dramas more feasible and popular.

We are not going to focus on the voice here, which deserves at least a book on its own, and neither are we going to focus on narratology and the construction and analysis of story. Rather we will consider how we can use sound to create and support story. While some nonfiction podcasts and audiobooks don't use sound beyond voice, the most engaging of them tend to use some form of sound design, even if it might be sparse in places.

In this chapter, we bring together everything we've learned so far: sounds, recording, space, effects, mixing, and meaning. We've already talked about many of the tools we have at our disposal, and we've practiced using many of these in our exercises. We've looked at how sound can help evoke sensory responses like imagery and touch, and now we're going to explore how we can use these and other techniques to support storytelling. This chapter will help us explore where, why, and how to use some of these tools in supporting a narrative.

9.1 Functions of Sound in Audio Story

To understand how sound can support story, it's useful to examine the functions or roles that sound plays in audio stories. Unlike film, television, or games, in audio stories sound is not supporting an *image* but instead plays a central role. Rebecca Parnell, a sound designer for games, explains what roles she sees sound playing in games specifically, but many of these also apply to sound design for any media:

> It can influence how the player experiences a game—how they feel. Do they feel tense? Do they feel scared? Do they feel happy? Does it feel fun? If it's an arcade game and it's high-adrenaline and you're scoring loads of points, how does the sound make the player feel? And that all comes in together, so it's not just about ticking boxes and, "Oh, this door opens and closes." How does it sound? Does it close softly? Is it slammed shut? What is the emotional impact behind that sound? And also another element, as well, especially with ambient sounds, is what is the context that you can give within the sounds? What information can you give to the player that you don't have to spell out in story and dialogue? How can you fill in the gaps? (quoted in Collins 2016, 97)

When it comes to audiovisual media, so much relies on visuals: facial expressions showing character emotion, camera edits and scenery to show scene changes, clear delineation of which character is talking and what role they play, what time of day it is, what the weather is like, what people are wearing, what era it is, and so on. When it comes to audio-only drama, sound has to fill all those roles. In a way, sound design for audio-only media is more complex, because a sound designer has so much more to convey without the image providing support. The many roles that sound plays tend to overlap, but I've separated them here so we can think about them in more detail and focus on each one.

9.1.1 Commercial and Aesthetic Functions

At its most basic level, sound plays a commercial role in some audio stories, including a sound signature or audio logo (*idents*, or *mnemonics*) at the start, any repeated interface or transition sounds that may be used to change or interrupt segments of a story (*bumpers* and *stingers*), and so on—these all have to do with branding the audio drama or podcast, helping to create a sense of familiarity with listeners to associate the work with a particular brand. When we have an episodic series, this branding aspect can be important to remind our listeners that they listened to (and hopefully enjoyed) a previous episode as they remember the sounds.

Related to the branding is the use of sound to improve our appreciation of the work, to make it more enjoyable, "cool," or indicate a particular genre or lifestyle. A podcast for hip-hop fans is going to sound very different, be paced differently, and use different mixing from a podcast designed about knitting for grannies, for instance. A newspaper ticker-tape sound is a classic news signifier; a gloomy but goofy evil laugh is a classic B-movie style horror signifier.

Exercise 9.1   Ident

Make an ident—a sonic signature, or audio logo—for a podcast series. Think about what you are trying to convey. Here are some suggestions to get you thinking:

An audio logo for a game company's podcast

An audio stinger for a new phone advertisement

An audio bumper for a quirky edgy news story podcast

Exercise 9.2   Audio Aesthetics

Find some sounds and music that would be suitable to accompany a podcast for horror fans, and then repeat the exercise for a podcast for car racing fans. What did you choose, and why? How does it relate to the brand the podcast might want to convey? What effects did you add to the sounds, and why?

Exercise 9.3   Thinking about Tropes

The classic news ticker sound and the evil laugh are common genre tropes. What other tropes do you associate with specific genres of story? Why do you think those tropes developed? Try turning a trope on its head—a horror news show, or a comedy horror show. What do you have to do to the trope to flip it?

9.1.2   Setting the Scene

With audio stories, we need to know fairly quickly where we are supposed to be located (in time and space), since we can't see the location the way we can in a film or game. Setting the scene or environment to represent a particular location geographically or temporally is an important function of sound. We've spent a lot of time listening to and thinking about our own soundscapes and the elements that go into our environment's sonic uniqueness. Can you think of the most important of those sounds? What shorthand signifier could you use to represent your particular location in space and time? What is the foreground, midground, and background, and how are these con-structed in the mix? Where are the sounds in space, and why are they placed there?

Environmental sounds can also be useful to indicate a change in narrative—such as a shift to a new scene. Some simple differences could indicate a change from day to night, for instance, by quietening down the ambience and adding crickets. Weather, time of day, geography, indoor/outdoor, season, ecosystem, and the like can all be sign-posted by using sound. Think in terms of foreground, midground, and background. What are the key sounds in the scene, and how do the midground and background support these?

Soundscapes also help create mood—they're not just *about* time and place but play a role in shaping that time and place as well. What does the ambience also say about the mood? Hard rain is different from soft rain, which is different from rain with wind. Adding reverb can put us at night, but also in a lonelier space. A soft wind in the desert is different from a howling wind in the desert. Adding distortion can make the sur-roundings sound more aggressive and closer to us. Think about how the ambience can tell us not only about the place, but also about how we are supposed to *feel* about that place.

Exercise 9.4   Impossible Soundscapes

Make an alien soundscape. What sounds did you choose, and why? What effects did you choose, and why? How would you change the soundscape from good alien world to bad alien world, and why? Make a future soundscape. What sounds did you choose, and why? How would you change the soundscape from good future (utopia) to bad future (dystopia)?

Exercise 9.5   Combining Place

Combine elements of two different soundscapes you've recorded into a new sound-scape to create a new place. Which elements did you choose to keep, and which did you throw away, and why?

9.1.3   Subjective Perspective

We looked at proxemics and the point of audition previously (chapter 3). Thinking about where you want the audience to be "located" in relation to the characters is important. Give them a perspective. Do you want it to be close and intimate, or are they at a distance? First person, third person, or omniscient? The perspective may change multiple times throughout a longer production. How will you alter the subjective position of the audience to enhance a story? What elements will you use to alter the subjective perspective? Sometimes, an audience is real or imagined in audio stories, for instance in the use of a laugh track. Listen to some audio stories and think about what the creators did in regard to their audience.

Exercise 9.6   Changing Perspective

Imagine we are in a scene where a serial killer is stalking someone. Shift the perspective from the victim's perspective to that of the killer and back. What did you do to shift the perspective, and why?

9.1.4   Structural Functions

Sound can be used to enhance or demarcate formal structure. Links, segues, and reverb tails can all signify a change in scene. For instance, tailing out the ambience of one scene and bringing in another scene with a hard cut can help your audience rapidly grasp a quick change of scene. Using a soundmark (for instance, a church bell) in the ambience can help the audience to know when we've cut back to that location after going somewhere else in the story. In this way, sound can be used to enhance continuity across scenes. For instance, if a character goes into a flashback, we might bring back the original ambience when the flashback is over to bring the listener back into the scene.

Sound can also be used to indicate a change in the narrative: if we've gone from a narrator's current day into the past and into the scene itself, for instance. And sound can be used for foreshadowing. Foreshadowing is a warning sign in advance of something happening at some point in the future. For example, the sound of thunder, even though the sky is clear, indicates that something bad is going to happen (i.e., a storm is coming, literally or metaphorically). Sound can be used to foreshadow events in story. But sound can also be used to set up an expectation related to a more immediate event. For example, the sound of a plane or bomb dropping from the sky sets up an expectation that we are going to hear a crash/explosion. We can use expectation to confuse the listener by disrupting that expectation. Foreshadowing, on the other hand, probably won't work if we disrupt it, because it's often only later that we recognize that the event was foreshadowed. Violating expectations, and then resolving that later, can provide an important sense of tension and release.

Another role related to the structure is the use of sound to pace the story, to take the audience on a journey. Changing the pace can create a sense of urgency or anxiety, for instance, or can be used to demarcate scenes. Listen to an audio story and think about how changes in scene are indicated in terms of sound. What types of transitions or segues are used, and how and why? How is silence used to create tension, emphasis, and focus?

Exercise 9.7   Structural Functions: Scene Segues

Use two very different soundscapes you've created or recorded. Try some different techniques to cut between the two as if they are two different scenes. What did you try, why, and what impact did it have?

9.1.5   Creating Characters

We learn so much about characters in audiovisual stories by their posture, their expression, their clothing choices or sense of style, and so on. None of those are available to us when we're just using sound, but we can still create a sense of who a character is. We can tell the difference between someone wearing a leather jacket and a ski jacket, for instance, by the sound their clothing makes. We get a sense of their gender by the way they walk and what type of shoes they wear, but we can also learn more about them by their footsteps: are they injured? Do they have a physical disability? Are they angry or in a good mood? There are all kinds of things just simple movements can tell us about characters. We can use *leitmotifs*—repeated sonic patterns, usually musical but not necessarily, that stand in for a character, to avoid having to use their name to indicate their presence in the story. The stereotypes and generalizations discussed in chapter 8 can come in handy, although we must also be careful not to offend. We can use panning or other spatial aspects to differentiate characters in the story if there are actors with similar voices, and place those characters into the space using the worldizing techniques discussed in chapter 4. Listen to some audio stories: How else are characters differentiated? How exaggerated are the sounds compared to real life? What are the characters feeling? Sound designer Walter Murch explains the use of sound to describe the inner life of a character in *The Godfather*:

> That example of the elevated train in *Godfather* is something that's primarily an emotional cue. There's rhythm to it but only to a certain extent, and story-wise it's a little ambiguous. "What is that sound? What is it doing in the film?" There's not an easy answer to that. But emotionally you absolutely understand what that sound is there for. Because there's nothing in the picture that is anything like a train—although reasonable that a train might be heard in that part of the Bronx—the emotion that comes along with that sound, which is a screeching effect as a train turns a difficult comer, gets immediately applied to Michael's state of mind.

Here is a person who is also screeching as he turns a difficult corner. This is the first time he is going to kill somebody face to face. He's doing what he said he would never do. (quoted in Jarrett and Murch 2000)

When it comes to nonhuman characters, sound design really gets to shine. Sound designer Stephan Schütze explained to me the complexities of designing creature sounds:

What I would try to do is, I'd try to look at a creature and not look at it and go, "Oh, it looks like a bear, therefore, I should use a bear sound." I'd try to look at it and think, what does that represent to me? Is it something that is a threat? Is it something that's a threat because it's a carnivore, or is it something that's a threat in the way an elephant is just simply because it's so big and it's potentially unpredictable? And often it's not going to run me down to eat me, but it might run me down to protect its own herd or to protect itself. But those are going to be very different sounds.

. . . So I think what you're trying to get, first of all, is you're trying to get what is the emotional impact you want that creature to have, and what is its behaviour like? How is it going to behave? Then you can start to think, alright, is it a big creature or a small creature? Now by and large, sound just works based on physics. A mouse is generally going to create a high-pitched and probably fairly quiet sound because it's got small vocal chords. It's like a short violin string, whereas an elephant, it has larger vocal chords, it has a larger set of lungs behind it. It has more bulk so it's likely to be a bigger sound. It's not always going to be the case, but it is something that as humans, emotionally, we respond to. If I build something that's supposed to be some giant Titan and it squeaks at you, unless I'm trying to purposely have some sort of comedic value to this thing, it's likely to go against what people are used to. So you need to sort of at least keep that in mind. So again, I've come to the creatures and I've decided what sort of emotional impact do I want to have.

Then it comes down to what sort of textural behaviour do I have, and I'll use the example of the dinosaurs. When I worked on the *Jurassic Park* game many years ago, I had to create the sounds of the dinosaurs. Now the obvious ones like the T-rex and the raptor, which you know from the movie, well, they were pretty much, we just used those sounds. But we had quite a large range of big herbivores. Now, the thing about herbivores, they're herd animals. So I had to work out their type of behaviour. Now, herd animals, generally you put one herd animal somewhere, they don't actually talk very much, because herd animals communicate. And so working out what types of communication they're going to have. Herd animals generally will have warnings to the rest of the herd, so there needs to be this kind of like, perhaps family, sort of like, "Hello, where is everybody else?" to, "Oh dear, there's something really, really bad about to happen, let's get out of here." And so then we've worked out the vocabulary, so I'd probably done a lot of this before, I've even started playing the sounds.

So then I get to the point of like, alright, well, what do I want to do? And I mention that I think bird sounds, pitching through those extremes, can be really quite amazing. And bird sounds, when you pitch shift them down a couple of octaves, or even more, you can get incredible textural sounds, and the thing is, I don't have to build. I don't have to artificially build a vocabulary. Because if I decide, for instance, we have a bird in Australia called the Peewee Magpie. It's like a small black-and-white bird. And they've got a range of sounds. But their range of sounds will be threat sounds, communication sounds. I've got all these sounds. The vocabulary exists, so I can pitch shift it down and I'm still going to get the sense of threat in a warning sound. I'm going to get the sense of community in a "Hello? Where are you?" sort of thing. I'm going to get the sense of urgency of "Oh, I'm hungry, I want to go eat" from the babies. And so in some ways, it's almost cheating, in some ways. It's kind of like I'm stealing another language and I'm just manipulating it to use for my particular thing. Now, I may need to chop and change some of those things, and that's the beauty. When I pitch shift them down, I might lengthen them, I might get longer phrases that I can then pull apart, and in a procedural way I can turn them into a language. (quoted in Collins 2016, 197)

Exercise 9.8   Creating Characters

Create a *nonverbal* dialogue between two characters, such as a woman and a man. You can pick the place and situation. How can you differentiate the two characters with sound (not using any voice at all)?

9.1.6   Believability and Immersion

Constructing a believable world is critical to creating a sense of immersion for your audience, allowing for or encouraging the suspension of disbelief, and adding realism. If sound designers were just mimicking the real world, sound design would be a lot less work. But we sometimes put sounds into scenes where they may not in reality belong, because their absence would make it feel *un*real. Nowhere is this more obvious than in representations of outer space, which of course is in a vacuum and so has no sound. I've previously called this *cinematic realism*—it's not realism, but it's what we've come to expect based on our past experience of screen media. It's believability, more than realism.

We've come to expect the "whooshing" sound of kung fu punches and kicks, the screeching of tires (even on gravel roads), the metallic "tshing" of a sword removed from a scabbard, and so on. Their absence would be noticed, and going for realistic punches might sound really boring for an audience used to hearing the dramatic hits of movies. Frogs make a variety of noises, but what sound do we associate with frogs? "ribbit." But the "ribbit" is actually just one specific species of frog located in Hollywood, called the Pacific Tree frog. A Hollywood movie needed the sound of frogs and sent their crew out to record frogs, and they came back with the Pacific Tree frog sound's "ribbit," which many people think all frogs should sound like, even if frogs in the rest of the world don't make that sound! The result is if you go out and record real frogs in the region where the program is set, the audience may not understand what the sound is. Again, it comes down to believability, not realism.

Physics can also be different in the auditory world. We know sound travels at a speed of 343 meters per second at 20°C in dry air. If we shot a gun toward a cliff one kilometer away, it should take nearly six seconds to return to us, but if we left a six-second gap to have an echo off that cliff in an audio story, the audience would likely be really confused. What we've become accustomed to based on our experiences of media can override reality, in other words, so understanding what conventions already exist—doing our research—is critical to believability.

Nick Wiswell, who designs the sound for many car racing games, explained the importance of finding the right balance of creativity and accuracy. As the explanation shows, a lot goes into all aspects of sound that play an important overall role in the audience's experience.

> Everybody has an opinion on the authenticity of a car sound. Some people have driven it in real life. Some people have seen it driving around in the streets or on a racetrack. Some people have watched a YouTube video. Some people have seen it in Hollywood, in a movie. None of those things sound the same. So, what is authenticity? Authenticity is your personal perception of how that car should sound. So we have to take all these things into account when we're trying to design. What is an authentic sound? Because you'll have people who will say, "You've got the sound of this car wrong. Look at this YouTube clip," that was recorded on somebody's cell phone that's just a big distorted mess. But they'll say, "This is what that car sounds like." And to them, that's their authenticity bar. Where somebody else will like, "I own this car, I drive it every day and it doesn't sound how it intended, how I remember it, or how I feel it should sound."
>
> All I know is I was there, or one of my team were there, to record the car. We have the recording of the car, so if I wasn't there at the session, my perception of the authenticity of that car is the recording we got back, but what if the recording we got doesn't capture something that people think of when they think of that car? So, one of my biggest jobs is to try and take all of these pieces and say, "Well, this *The Fast and the Furious* movie made the car sound *this* way, but this YouTube clip says it sounds like *this*, which is completely different. I've got this recording, that doesn't sound like either of those two. There's a guy in the office who used to own this car, and he says, 'This is how it should sound.'" So all of them are right, and none of them are right all at the same time. (quoted in Collins 2016, 306–308)

9.1.7 Connotation and Metaphor

Sounds can have many connotations, and sound effects can be used in subtle or quite obvious ways to fill in the blanks in what is said, or add slant or bias. For instance, we could use a politician saying "There is absolutely no risk of war" and have the sound of guns and bombs in the background to give the impression that we think the politician is lying. The sound can stand in for something else and through this relationship reveal something about character, plot, theme, product, and so on, as *metaphor*. A metaphor, if you've forgotten your high school English class, is a figure of speech in which a phrase (in our case, a sound) stands in for or is symbolic of something else.

Walter Murch has written a great article on sound and metaphor in film, and while most of it applies to sound and image, one great quote stands out: "The metaphoric use of sound is one of the most fruitful, flexible and inexpensive means: by choosing carefully what to eliminate, and then adding back sounds that seem at first hearing to be somewhat at odds with the accompanying image, the filmmaker can open up a perceptual vacuum into which the mind of the audience must inevitably rush" (Murch 2000). The use of sound on its own, without image, can serve an equally important role as metaphor. A rising sequence of frequencies has different associations from a falling frequency: we could use rising frequencies to represent a training sequence in a story about a boxer, or falling frequencies to represent that boxer's fall to the mat after a knock out.

Exercise 9.9   Sonic Metaphors (Partner/Group Exercise)

Come up with a list of sonic metaphors, and compare them with friends if you can. Are you "on the same page" when it comes to metaphors?

The juxtaposition of sounds with each other can lead to a form of *closure*. A concept from Gestalt psychology, the idea of closure is that our mind fills in the blanks when given certain parts of a whole. We could call the phantom fundamental (section 2.5.1) a form of closure, since our brain is filling in the blank of the fundamental by using the harmonics of the tone. But we could expand the concept of closure to include how the brain can fill in information that isn't there to provide a sense of completion. We may fill in the auditory ambience that gets ducked to focus on other elements of a sound-scape, for instance. In the last chapter, we touched on intersensory integration, and it's often suggested that "the best pictures are in the head": part of closure may involve filling in the visual blank of radio listening with sound. As Nick Wiswell describes above, we don't have to use all the sounds in a car to represent a car—we can just use a few and our minds will fill in the rest.

We can also use *hyperbole*—the obvious exaggeration of a sound to make a point, counterpoint and contrast for irony or humor. We can use *anthropomorphism* to enhance a listener's feelings toward or empathy with a nonhuman character (think *Wall-E*).

Exercise 9.10   News Story in Sound

Take a news story from the day and tell its story in just sound: how might you use sound in some of the metaphoric or ironic ways just mentioned to emphasize or criticize the journalist's point of view?

Exercise 9.11   Anthropomorphic Sound

Grab a stuffed teddy, robot, creature, action figure, or nonhuman doll from a junk store, and give it a series of sounds. Imagine animating this figure: how would you anthropomorphize the sound of it, and why?

9.1.8   Mood and Emotion

There is a saying in the video game industry, by composer Marty O'Donnell: "Music makes you feel, sound makes it real." I disagree with Marty, because sounds can make you feel as well! In fact, sounds can express emotion, underline emotion, as well as induce emotion: by this distinction, I mean it's possible for sound to tell us how a character is feeling without necessarily making us feel that way, but it's also possible to make us feel that way. We can hear a character is angry without feeling angry ourselves; but some sounds can actually induce a particular emotion in us.

We looked at some effects and talked a bit about how reverberation can feel warm, phasing can feel psychedelic or mimic the effect of drugs or mental disturbance, overdrive can feel warm but when pushed can feel more like anguish/anger. It's not just the choice of sound and the effects we use; as we have discovered through our journey, other aspects of sound impact feeling. The proximity, loudness, and directionality of a sound will also have some impact on its affect (see Tajadura-Jiménez et al. 2010). The context in which the sound is heard will also influence the perception of the sound. It is extremely difficult, then, for us to determine the effect of a sound on every person in every context, but we work with generalities and with our intuition that we develop over time and practice.

Joanna Orland, a game sound designer, describes how the selection of sound in the ambience can influence emotions:

> I believe that audio plays a huge role in the emotional effect on the player in the game. Audio is what is manipulating them emotionally. We can use almost psychological tricks to ensure that they are feeling something without them even noticing because when audio is done well, no one notices it. It's only when it's done wrong or poorly that people take notice. I believe, with audio, if your audio is well thought through, you can use emotional tricks. So for example,

with the horror genre, there's quite an obvious pattern that film does with the soundscapes, you know, the environments are always very abstract and hyper-real, which is an exaggerated form of reality, and the character sounds are always very naturalistic and realistic. And I believe it's a psychological trick because the player, if it's a horror game, or the viewer, if it's a horror film, can feel the emotional attachment to the character, because they recognize the sounds as human and as fragile, almost because it's very naturalistic, and almost frail, in their human elements. But then when this character is placed in this really hyper-real environment, there's a tension that's created because this character that they believe in, because they've got an attachment to them, is put somewhere very unfamiliar, which is scary and unnerving because this environment sounds really alien, and they don't know what can happen to this character which they've got an emotional bond with. (quoted in Collins 2016, 94)

Emotions are highly complex, and the use of "mood maps" to pigeonhole emotion is controversial; however, mood maps can be an effective tool to help us to define general categories for discussing sound's affect and think about our use of emotion in sound design. There have been similar attempts to categorize and label emotions, such as Plutchik's "wheel of emotions." A mood wheel can be useful to think about what characters are feeling, and what you want the audience to feel at any point in the story. Plutchik's wheel, shown in figure 9.1, puts emotions on spokes that radiate outward from most to least intense (see, e.g., Donaldson 2017). As you progress in your sound design skills, you'll come to associate emotions with your own bag of tricks that helps you to convey that emotion.

Plotting the key emotional points in the story and aligning scenes with moods can help you think about how to emphasize key moods. We talked a bit before about using dynamic range for emphasis. What are the peak emotions, and what are the peak sounds? At what point do they appear in the story, and how do they ramp up? How does dynamic range draw the audience's attention toward that particular event? (As discussed above, dynamic range and amplitude are not the only ways to show emotion, of course.)

Exercise 9.12   Sonic Mood Board

A *mood board* is a common design tool where a collection of colors, textures, and images are gathered to represent a particular mood that a designer is trying to explore for a product or client. For instance, if I were trying to create the feeling of serenity or calm I might use images of calm lakes, being snuggled inside an igloo, a candle, a clear sky, soft fluffy textures, and so on. We can create mood boards with sound, too. Find sounds that represent a certain mood because of their signification,

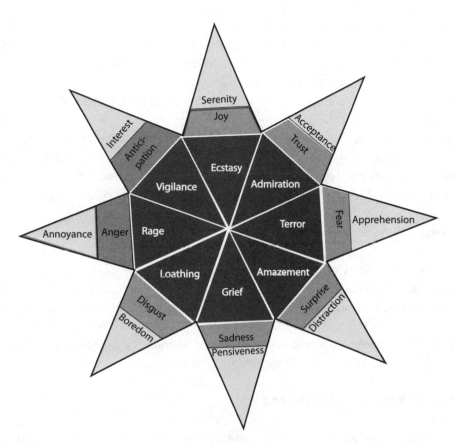

Figure 9.1
Plutchik's wheel of emotions (based on Plutchik 1991).

their associations, their textures, and so on. Don't use any vocalizations (no voice, no "emotes," or vocal utterances), and no images are allowed. Use effects to enhance or alter sounds to really bring out the mood. It's fairly easy to choose strong emotions like anger, but how would you sonically represent something like "wistful" or "pensive"?

Exercise 9.13   Tension and Release

Create a new mood composition. This time, focus on building up tension for the first part of the piece, then give the listener some release at the end.

9.1.9 Information, Feedback, and Reward

Finally, in some types of audio stories (interactive stories, for instance), sound may play an important role in providing some form of feedback to the listener—they clicked on the right button, they collected a good or bad object, and so on. Providing some information about what is going on in the story, and giving feedback to a user in any kind of interactive story, can be one of the most important things that sound can do. This feedback can also come in the form of some form of sonic reward for performing a correct action. In an audio-only game or interactive story, for instance, undertaking a particular action that results in winning or scoring points should be accompanied by some auditory reward. This "audio bling," as I like to call it, provides positive reinforcement.

Exercise 9.14   Sonic Storyboard

Create a sonic story of about 30 seconds using only sound (no music or dialogue). Document the effects and techniques you used. Some examples of the types of stories you might describe include: 1. You arrive home and scare your cat or dog, who knocks over your favorite vase. 2. It's the bottom of the ninth inning, and you're up to bat. 3. Alone in a parking garage, you fear someone is following you. 4. You are lost in the woods, and spend the night in an abandoned barn.

Exercise 9.15   Sonic Storyboard 2

Take your sonic storyboard. Now assume that the entire story takes place in a dream or hallucination. How will you change the sound?

Exercise 9.16   Sonic Storyboard 3

Flashback or flashforward from your story to create the next scene. How will you show the perception of time passing (hint: don't use a clock ticking!). How did you transition from the first scene to the next?

9.2   The Mix

It's important to note that although we're not talking about voice or music in this book, these audio aspects may fall under the purview of a sound designer's job on an audio story team. A sound designer may be responsible for dialogue recording and/or editing, and selecting or commissioning music. Even beyond that, the sound designer may be responsible for thinking about all of the other aspects of dealing with dialogue and music in the mix: How are sounds interwoven with dialogue and music? Are they given their own space? How do sounds or ambience frequencies interact with dialogue and music—have the sounds been selected or mixed to sit in a different acoustic space? Are there places where there is no distinction between the music and sound—that is, can we use sound design as music or in place of music? Where, why, and how?

Regardless of what your overall role is on the team, if you're responsible for sound design, you need to be thinking about the music and dialogue and how your sound will fit with it. How is ducking used, and where? How might you use EQ instead of ducking to achieve the same aims? How else might you deal with mixing the audio components together? How are the files being distributed and how is your audience listening to the material? All of the elements of mixing need to come together, because great sound design can be ruined by a bad max.

9.3   Audio Research

We touched on reference material in relation to mixing in chapter 6 and above. Reference materials are the examples from media that we may first turn to when researching a project. If we are assigned the task of creating the soundscape for an alien planet, one of the first things as a sound designer that we might do is see what other people have done in the past. We may watch movies, read books that describe the sound of distant planets, play video games set on distant worlds, and so on. While it's certainly possible to try to create an entirely new take on alien planet soundscapes, we know there are some conventions that audiences rely on (see above). Understanding how those conventions have been used in the past is key to understanding how much we can push our own design. Joanna Orland, a sound designer for games, explains:

> What I do is, I start off by creating an audio vision, which is what the audio's goal is for the game that I'm working on. For example, when I worked on *Wonderbook: Book of Spells*, that was a magic game, so I had to figure out what the sound of magic was. So I did a lot of research, I watched a lot of films, read some books, and I discovered a pattern in mostly the film stuff that I watched. Basically, in a lot of films and popular culture, the magic sounds are sourced from nature. So I decided for *Book of Spells* that was what we were going to do as well. So I created this whole audio vision that the sound of magic would be sourced from nature. And then, as a sound designer, I had to then create these sounds that fi t into this world that was being sourced from nature. I think we do look to film quite a bit as an example of how our audio should sound, because the mainstream is pop culture, and film is a popular media, and that's what people are used to, these conventions that we hear are defined by the films that we watch. So with games, we don't want to do something so unfamiliar to people. We want them

to believe in our vision, so a lot of the time we will use familiar tropes and sounds and methods that film may use. . . . It's effective, it's proven, it works. So we're not reinventing the wheel in games. (quoted in Collins 2016, 93)

When films are being made, they will often use *temp tracks* for music—music that is the feel or atmosphere that the director is aiming for. That often gets sent to the composers so that they have an idea of the type of music the director wants. Sometimes, the temp tracks even end up staying in the film, if the director or others form an attachment to seeing the music. When it comes to sound, it would be very rare to get temp track ambience or other sound design, but we can still think along those lines: that there is a template that exists for what we want to do. We may not always want to rely on the template—after all, we want to flex our creative muscles—but taking a risk on a completely new approach isn't always our job as sound designers. We want to evoke a certain location or mood quickly and effectively. Most of the time, we actually don't want to be *noticed* (other than by other sound designers!), so we can't stick out too much.

9.4 Audio Story Analysis

Analyzing how other people are effectively (or ineffectively) using sound is one of the best ways to explore sound design for story and do your research. The more you train yourself to be thinking about sound design, the better equipped you'll be to approach your own sound design. Below is a list of just *some* of the questions you should be asking: and not all questions will apply to every audio story.

Functions: What functions does the sound play in the audio drama at different points, and why?

Sound selection: How does genre affect sound design choices? Consider their choice of key sound effects: Why are these the most important sounds? What sounds are used, where, and why? How are characters delineated by sound? What sounds are part of the branding or signature, and why did they choose those sounds? What tropes or stereotypes are inherent in the sounds chosen? Which sounds are symbolic or metaphoric, and what do they signify? Which sounds are universal and which are conventional? What sounds are used (or emphasized) for believability, rather than realism? Do any sounds anthropomorphize objects, or serve as a character in themselves? Which sounds evoke a bodily response, and why? Which sounds evoke a haptic or visual response? What actions are heard in the sounds?

Effects (DSPs): What effects are used, where and why? What impact do the effects have on the listener and why? As we have touched on to some extent, sound effects can alter the feeling of a particular sound. How are flashbacks (if any), or in-head thoughts and feelings, created sonically through effects?

Mix: How do foreground, midground, and background change in the mix (if at all)? Are mixing choices realistic or creative? Why do you think they chose to mix it that way? How is the sound mixed in relation to music and dialogue? How is dynamic range used to emphasize certain sounds over others? What is the most important scene, and how is that shown in sound?

Structure: How are scene changes shown in sound? How do they build tension? What sounds carry between scenes and why? What sounds are specific to scenes, and why?

Spatialization: Is it in stereo, binaural audio, or some other format? Why do you think they chose this format? What tools are used to spatialize sound in a way that supports the story? Consider the different points of audition: How does the sound help us to empathize with characters through point of audition? What techniques support the point of audition?

Beyond these more technical questions are the personal ones—what did you like about it, and why? What would you have done differently?

Exercise 9.17   Audio Story Analysis

Get cracking on your own analysis. Remember you're going to have to listen to it several times for different things as you go through it, so pick one that isn't too long.

Exercise 9.18   Comparing Approach

Listen to two popular podcasts or radio dramas and compare their approaches to sound.

9.5   Spotting a Script

Many writers describe sounds in their stories in order to evoke sounds in their readers' minds. Whether you are recording a fiction or nonfiction audio story, understanding where, why, and how to use sound is critical to its success. Scriptwriters sometimes write some key sounds directly into scripts. For instance, here is one part of the script for BBC radio series *The Wire* for the episode "The Startling Truths of Old World Sparrows" by Fiona Evans (2013, 11):

```
STAN'S KITCHEN. DAY.

   DOG WHINES.
                            STAN
   What's the matter baby? Do you want to go out? She doesn't
like the snow, especially when it's deep. With her being so
small. Poor thing . . .
   STAN UNLOCKS AND OPENS THE BACK DOOR
   Her belly gets wet. Go on, go out for a wee.
   KIDS LAUGHING. A WHEELY BIN IS KICKED OVER. THE KIDS CALL HIM
NAMES
   STAN GOES OUTSIDE. SHOUTING AFTER THEM.
   KIDS RUN OFF.
   Oi! What do you think you're doing? Oi . . . Don't you come
near my house again, do you hear me?
   Bloody snow! Can't run in this can I?
   What shall I do? I'll clear the path. That way I can run after
'em next time.
```

Note how the scene is delineated by all CAPITALS, BOLD AND UNDERLINED. The character is delineated with an UNDERLINE AND ALL CAPS. Dialogue is not bold, all capitalized, or underlined. ACTION IS ALL CAPS.

In this single scene, we have two locations, which is unusual, although technically the scene is inside Stan's house, and just in the doorway/just stepped to the outside, but we need two ambiences to delineate the space: inside and outside, even though it's one scene. Some writers may have separated those scenes out into two distinct locations, or they may be thinking like a camera person and imagining the entire scene is shot from the kitchen, with some view through the doorway.

We have several characters, even though only one has dialogue: Stan, the dog, and the kids. We have some action events: dog whining, door being unlocked opened, wheely bin being kicked (a wheely bin is a garbage or recycling container on wheels). These are the key spot sounds, or foreground sounds, in other words. But so many other sounds are also a part of the scene that we need to include: footsteps, wind or birds outside (depending on time of day and weather, based on context from the rest of the script), kitchen sounds, and the like.

Now we can think in terms of adding emotional impact or the things that are not there. Let's suppose Stan is talking to himself, not to someone else. What might his home sound like? Maybe the faucet drips because he's not attentive to those types of things. Maybe he stumbles a little and kicks the dog bowl as he walks over to the door. Maybe the door is creaky and old-sounding. Maybe the lock is not one simple lock but has several stages that need to be unlocked. Maybe the door has a draft stopper that drags on the floor as he opens it. Maybe the floor of the kitchen is wooden and creaky, or maybe it's tiled and hard. Maybe it's blowing a gale outside, or maybe it's a cold crisp night. What about the kids? Are they all boys? All girls? What approximate age—little children sound different from young teenagers. What's in the wheely bin as it's tipped over—paper recycling is going to sound different from glass or garbage. These are the creative decisions we have to make that can relate to the mood we want to convey.

When you are spotting a script, during the first read, it helps to read the whole script, paying attention to objects, actions, environments, emotions, and so on. Circle keywords that indicate where and what sounds are linked to specific characters, objects, or actions. Think about the tension and release points. Write down your thoughts as you read.

Exercise 9.19   First Spot

Find an audio drama script (there are many online scripts available: make sure it's an audio drama, not a television or film script). Read through it and circle the keywords that would help your sound design. In this case, it's the location, the characters, and the key foreground sounds.

On the second read-through you can think more thoroughly about what sound can bring to the story. What is the overall theme of the story, and what sounds do you associate with that theme? What sonic generalizations or stereotypes do you associate with the time and place, the story, the emotions, the characters, the key objects? Can you adjust your sound effects to enhance the mood or emotion of the key sounds you've circled or added that might convey story? Using a mood wheel, mark off which parts of the script associate with which moods. What is the overall arc of the emotion, and how can you convey that with sound? What is the peak emotional point and how can you convey that? How do places and times shift in the script, and how can you convey that with sound? What is the overall genre, and what sounds do you associate with that genre? If Stan is in a horror movie, for example, that's a different sound than if Stan is in a comedy.

Go through the questions presented in the analysis above, and consider how you might apply what you learned from analyzing other works to your own work here.

Exercise 9.20   Second Spot

Conduct a second (and third and fourth!) read-through on your script, marking out the key aspects you can consider with relation to sound design.

9.6   Cue Sheets

After the first spot, we can make a *cue sheet* for the script. A cue refers to when a director would point to a person in live radio, recording, or film, indicating whose cue was coming up, so it distinguishes sound, music, and dialogue. We should have a separate cue for each distinct music, dialogue, or sound event. Don't forget we need time in between dialogue to fulfill the functions of sound in the audio drama, and don't forget to include silence where appropriate. Here is how we might approach the previous scene by adding cues (cues are usually numbered in a cue sheet, as well):

1. MUSIC: (bridge) TRACK 13—FADE UNDER
2. SFX: DOG WHINES. SCRATCHES AT DOOR.
3. STAN: What's the matter baby? Do you want to go out? She doesn't like the snow, especially when it's deep. With her being so small. Poor thing . . .
4. SFX: STAN FOOTSTEPS OVER CREAKY FLOORS PAST DRIPPING TAP AND UNLOCKS AND OPENS THE DEADBOLT AND SLIDING LOCK.
5. MUSIC: TRACK 14 STING
6. STAN: Her belly gets wet. Go on, go out for a wee.
7. SFX: DOG FOOTSTEPS IN SNOW. KIDS LAUGHING. A WHEELY BIN IS KICKED OVER. BROKEN GLASS. THE KIDS CALL STAN NAMES
8. SFX: STAN WALKS OUTSIDE INTO SNOW. SHOUTING AFTER THEM. KIDS RUN OFF RAPID FOOTSTEPS MUFFLED BY SNOW. HOWLING WIND
9. STAN: Oi! What do you think you're doing? Oi . . . Don't you come near my house again, do you hear me?

    Bloody snow! Can't run in this, can I?

    What shall I do? I'll clear the path. That way I can run after 'em next time.
10. MUSIC: TRACK 15 FADE IN BED

In the cue sheet, music and sound are underlined. The whole thing may be boldface (these are based on conventions from hand-typing things out and relying on a sometimes poor photocopier quality in the old days), so underlining differentiates dialogue from sounds. You'll note that not all sound effects were included. I also split the sound effects up in cue 7 and 8, because these are two quite different cues. If Stan were to cower and run inside, we might approach cue 7 differently. I also didn't list all the sounds we discussed above, but I included a few that weren't in the script to remind me. We can fill in the blanks on those as we go.

When it comes to music cues, these may or may not be up to the director, the composer, or to you as the sound designer. Let's talk about music cues.

We started with a bridge: a bridge is a music cue that plays between scenes without dialogue. Sometimes it's referred to as *act in* or *act out*, as it brings the actors in or out of a stage in a theater production. We also had a *fade in bed*: we know what a fade in is, and a *bed* is music that plays under dialogue or action. These are sometimes listed as *source bed* (in the *diegesis*, or story-world that the character hears, such as a radio on in their home), or just *music bed*, which is composed music that the characters don't hear (nondiegetic). Fade ins differ from *fade unders*, in that a fade in finishes before dialogue or sound effects start, and fade unders don't finish until the other sounds have already started.

We also had a *sting*: this is sometimes called a stinger or bumper. In the old days, when something dramatic was going to happen, the organist would play one or two notes repeatedly to underscore important dialogue and create tension. It's also used to refer to a quick "blam!" music chord or clip that draws attention quickly then gets out. Bumpers tend to refer to the music specifically used to start a scene or break for commercial, but in a drama, a stinger tends to be a dramatic moment, and may signal the start of a new scene.

Some other music cues we haven't used here:

Under: Leave music playing at same volume as other sounds happen.

Duck under: duck the music under what is happening next.

Establish: let the cue play for a bit to establish the music/scene.

Quietly in B.G. (or just B.G.): playing in the background.

Self-fade: the composer has designed the cue to fade out at the right time.

Play through and out: like a self-fade, the cue is designed to be played in full.

There are other things as a sound designer that we may want to add to the cue sheet that weren't in this scene. If there is a walkie-talkie, or talking from behind a door, we may want to add [FILTER] or [REVERB] to the dialogue line. We may want to also add any emotes—any brief vocalizations, like [GASPS]—as Stan opens the door.

When it comes to sound effects, there are a few things you may encounter that we haven't talked about before: *walla-walla*, or just *walla*, is the term for crowd mumbling, as you would hear in a busy place like a train station, restaurant, and the like. It's part of the ambience but sets the scene as being busy with people. A *ramp* is a music cue that leads up to vocals.

In addition to a cue sheet that may be written by someone else, we may get production notes: "Cue 4 on Page 8 should be all in heavy reverb," or some such note. It is our job as the sound designer to either follow the direction or make a case for why the direction isn't adequate for our vision, and, in that case, we should record several versions to show the director what we mean and why it's superior to their vision.

Exercise 9.21   Creating a Cue Sheet

Using the script you used in the previous exercise, go through and create the cue sheet for the entire script. You will need some time to really think about what you're doing, so don't rush it.

9.7   The Asset List

Once we have a sense of what sounds we're going to need, the next step is to create our asset list—a list of all the sounds we need to find, record, make, design, and so on. How we divide or categorize these is up to us, unless we have a producer who has already decided what they need to share with a team of people. We may, for instance, create lists of sounds we know we're going to have to rely on a sound library for, and sounds we know we can record ourselves. We may also divide them up by scene, or by type of sound (spot, ambience, etc.). Asset lists are typically spreadsheets that help us to keep track of what sounds we need and whether we've got them ready to go into production. Asset lists are also great places to start a naming convention for our project. If we know we need "Josie's footsteps," then when we slate and name our files, we can name the files something like josies_footsteps_indoor.wav, or Josies_footsteps_02.wav, and so on. Spreadsheets can also be color coded, filtered, and so on, so we can use them to track which sounds are ready, which are rough and need redoing, which are not yet finished, and so on. The conventions we choose are up to us (and our team), but developing and sticking to one convention is really useful. An asset sheet might look something like table 9.1.

Table 9.1 Sample asset list

Scene	Sound	Filename	Category	Notes	Finished?
Fireplace	Fire crackle	Fire_01.wav	Ambience	Loop. Still needs reverb.	—
Fireplace	Dog whimper	Dog_02.wav	Spot	Get John to voice this.	—

Exercise 9.22   Asset List

Grab your cue sheet from exercise 9.20 and create an asset list for the project.

Exercise 9.23   A Writer's Description

Ray Bradbury's story "The Foghorn" describes the sound of foghorns (1983, 434–436): "I'll make a voice like all of time and all of the fog that ever was; I'll make a voice that is like an empty bed beside you all night long . . . like trees in autumn with no leaves . . . a sound like November wind and the sea on the hard, cold shore. I'll make a sound that's so alone that no one can miss it, that whoever hears it will weep in their souls . . . and whoever hears it will know the sadness of eternity and the briefness of life. . . . The Fog Horn blew. And the monster answered. A cry came across a million years of water and mist. A cry so anguished and alone that it shuddered in my head and my body. . . . Lonely and vast and far away. The sound of isolation, a viewless sea, a cold night, apartness. That was the sound."

Make the sound of this foghorn based on his description. If you can think of other sounds that are described in books you've read, try to create those sounds too. In many cases, we may not be able to talk to the writer to ask them what they were thinking; we can only go by their written descriptions and our own imagination.

Exercise 9.24   Recording a Script: Radio Drama/Fiction Podcast

Take your final cue sheet and record it. If you can't find actors, you can always read it yourself. We're just focusing on the sound effects, so the voice is less important here. What did you find in recording your script that you needed to change from your initial spotting session, and why? What did you add/subtract?

Reading and Listening Guide

Clive Cazeaux, "Phenomenology and Radio Drama" (2005)

Cazeaux argues against the idea that by not having image radio drama is an incomplete or "blind" artform. Drawing on phenomenology, Cazeaux shows how sound design in radio drama offers us a complete experience by using the way intersensory perception helps sound to evoke images and touch.

Neil Verma, *Theater of the Mind: Imagination, Aesthetics and American Radio Drama* (2012)

Verma's book covers a history of radio drama in the US from a variety of perspectives, including a particularly in-depth exploration of subjectivity and the point of audition (a term he takes issue with). Verma gets quite detailed in descriptions of studio set-ups and microphone techniques used by radio, and overall the book is very useful for another perspective on sound recording for radio, particularly in ways you might analyze and think about radio.

BBC Academy Podcast: "Working with Sound" (May 25, 2017), and "How to Create Stories with Sound" (March 23, 2017)

A BBC podcast series about creating podcasts: a few episodes are specifically about sound and have clips of interviews with some of their sound designers.

Rob Rosenthal, *HowSound*

A Transom.org podcast about radio storytelling, and many of its episodes are about recording and sound design for podcasts: https://transom.org/2005/walter-murch/.

Radio Drama Revival

A biweekly podcast that's been going for over a decade now. Each episode discusses the craft of radio drama, but also presents dramas and interviews the people involved in creating them.

10   Conclusions and Wrap-Up

We've covered a lot of material about sound design in this book. We've learned a variety of listening practices and have begun to train our ears. *Begun*, you ask? Yes, just begun! After two decades, I still feel as if I've only just begun. The deeper I dig into sound design, the more layers I uncover. For each chapter in this book, I've really only scratched the surface of what is possible. I encourage you to seek out additional resources and dig deeper yourself, and to keep practicing.

We started out by learning how to listen. This is a lifelong practice that gets easier over time. As I mentioned, though, it comes with a particular affliction: once we've learned to turn our ears on, it's very hard to turn them off, and we have to learn to live life with a sensitivity to sound in a very noisy world.

We learned about the acoustic properties of sound, how to describe and edit these properties, and how sound-processing effects can mimic some of these properties. We also learned the basics of recording and mixing, but there's an awful lot we didn't cover when it comes to different types of recording, and various practices from field recording, studio, film, games, Foley, and music recording. You'll develop your own preferences and practices over time, but I encourage you to record as much as you can, listen to those recordings with fresh ears the following day, and always continue to take notes in your journal.

We began to explore the creative uses of sound, and this, too, is something that you will develop further over time and through practice. Every project has its own needs and you may find yourself spending days or even weeks just looking for that perfect sound for a character. You'll also find you may pick up another habit, one that others find particularly annoying—banging, squeezing, and touching everything you encounter to hear its properties. I used to carry chopsticks or a pen around with me so I could tap various surfaces and listen to the sounds they made. I no longer do this, but I do pick up, shake, tap, and touch just about everything that is new to me and make a mental note of these objects. New spaces offer some really exciting experiences, and many places on the planet are worth exploring for their acoustic properties alone.

We didn't cover a lot of material about the context of sound design, other than with storytelling. I mentioned that most sound design books are about sound for image, and certainly that's one important context. You'll find that when you put sound design to image, some of the aspects of sound are perceptually altered, and what worked great on its own no longer works with image. There are many books out there about sound and image, and most sound designers work to image, so it's important to also develop that skill. Even before we get to image, the overall sound composition, or soundscape, changes the context of sounds, as you've probably witnessed in some of your compositions. As Walter Murch put it, "The chemistry of soundscapes is mysterious and not easy to predict in advance" (quoted in Ondaatje 2002, 245).

The soundscape is of less concern to product sound designers, although complex products require careful consideration, as does how the product will mix with existing environmental sounds: we have a vibrate function on our cell phones, for instance, because sometimes a ringtone is inappropriate. Some products are built to mask other sounds—white noise generators, for instance, are often used by people sensitive to environmental sounds when they sleep, or to mask conversations in a busy office. Other sounds (a smoke alarm, for instance) we need to hear over all existing environmental sounds. In other words, the selected sounds all affect and are affected by their context, and sound designers must make selections in terms of the acoustic effects of the sounds (to cut above masking, to avoid certain frequency ranges, and so on), but also in terms of the mix—what is emphasized, and what is removed.

Other contexts also must be taken into consideration. The location of listening and the playback technologies also influence design choices. A movie played in a full surround-sound theater will sound very different from one viewed on an iPad with headphones (or, worse, using the iPad speakers). With podcasts and radio dramas, we can assume that many listeners will be listening on portable devices and potentially competing with many external noises—the train the listener is on, the buzz of other people on machines at the gym, the hum of the car's wheels, and so on.

There are also the wider contexts of genre, style, and brand. Increasingly, branding affects the sound design for products of all kinds. Sound designers must understand the semiotics of the sounds that they create, which is to say, what the sound is communicating in terms of the brand. Mercedes, for instance, introduced the visual Mercedes Benz logo, created by Gottlieb Daimler, in 1909, but it was not until 2007 that they introduced their sonic logo. As part of a larger brand redesign, the acoustic trademark, according to Dr. Olaf Göttgens, Vice President of Brand Communications for Mercedes-Benz Cars, is "going to make the Mercedes-Benz brand not only visually, but also acoustically distinctive, and thus more quickly recognizable. This acoustic trademark is a perfect fit for Mercedes-Benz—it is emotional, elegant, and unmistakably associated with our brand" (Mercedes 2007).

We also didn't cover the process of developing sound design as part of a team. Whether we're doing a radio play or podcast, a theater production, a film, a game, or a product design, we're likely going to have to work with others on the project. Communication is a key skill that plays an important role in design. There will be times when we have to advocate for sound, because others won't immediately see its value. At other times we'll have to understand that the music may need to take precedence in the mix, and we must cede control of key sounds that we've spent a lot of time creating. There may be times when we are given complete autonomy, but more likely others will bring their own visions to the table and they may conflict with our own. We may have to work on projects that don't seem immediately interesting for sound, and we may have to find our own way into the material to create that interest. We may spend all our time creating a masterful mix of beautiful sonic wonder, only to find it has to be compressed down to a small size and sounds (to our trained ears) terrible in the final product.

These are just a few parameters of sound design that we didn't cover in this book—or just touched on. You'll find more as you progress in your journey.

Exercise 10.1   Extra Credit 1: Sound Design on a Time Budget

You have forty-eight hours. You have to design the sound for a nonfiction radio program about Three Mile Island. You haven't been given any other direction, script, or recordings to support the design. Gather the sounds and music you might need, and create some ambiences.

Exercise 10.2   Extra Credit 2: Sound Design Pitch

You have twenty-four hours to prepare to pitch a sound design for a historical dramatization about the American civil war. Bring together everything you've learned to describe how you would approach the project. Remember to describe it in a way that really sells your ideas about the project.

10.1   Self-Evaluation

The journey into understanding and using sound in design practice is not something that can be taught effectively in a few months. By keeping a journal, though, and reflecting back on how far you've come, you will realize how much you've learned. You can repeat many of these exercises multiple times as you train your ears, and compare them to earlier versions you created. I encourage you to continue writing in your journal and keep a notepad handy for reflecting on your practice. Not only will you see how far you've come, you'll think of things to share and ways to help teach the next generation of sound designers.

Exercise 10.3   Reflection

Reread your journal from start to finish. What have you learned since you began? Make a list of all of the things you learned. How has your position as a listener changed through repeated listening practice? Repeat some of the earlier exercises now that you've trained your ears. What are the differences you hear now in your progress?

Exercise 10.4   Kill Your Darlings

The writer William Faulkner once gave the advice to "kill your darlings." The advice refers to the overuse of something that becomes your cherished darling. It might be an effect that you think makes your sound stand out from others. It might be a particular approach you've taken and repeated multiple times in the exercises. You may not have been able to get the distance you need to reflect on what your darlings are, so you may need to seek out the advice of someone else in sound design. Go back through what you've done and kill off a darling or two, and then repeat the exercise without those crutches.

Exercise 10.5   Simplify, Simplify

A common mistake in all design practice when someone starts out is to get too complex, to show off the skills you've learned. How many times have we heard someone say of a seemingly simple design, "I could have done that"? When we don't understand that sometimes the simplest designs can take the longest, it's easy to dismiss a design as "easy." I can remember thinking that of Ducati motorcycles' logo change in 2008: it became a simple circle with a line through it. Who couldn't design that in five minutes with a computer? It's deceptively simple, but the more you think about it, the more complex you realize it is in its simplicity. Take one of your sound designs and make it deceptively simple. How long does it take to get the same idea across with less? Probably a lot longer than you thought! Simplification is a sign of mastery.

Fortunately, many wonderful resources are available today for learning about how others approach their sound design practice. As a final project, why not start a resource yourself? Create a podcast about sound design, or teach others to use the tools you've learned now.

References

Aizenberg, Mark, and Maria N. Geffen. 2013. "Bidirectional Effects of Aversive Learning on Perceptual Acuity Are Mediated by the Sensory Cortex." *Nature Neuroscience* 16, no. 8: 994–996.

American Institute of Physics: Inside Science News Service. 2000. "The Secret of a Tiger's Roar." *ScienceDaily*, December 29. http://www.sciencedaily.com/releases/2000/12/001201152406.htm.

Augoyard, Jean-Francois, and Henry Torgue. 2009. *Sonic Experience: A Guide to Everyday Sounds*. Montreal: McGill-Queens University Press.

Bradbury, Ray. 1983. "The Foghorn." In *The Stories of Ray Bradbury*, volume 1, 432–440. London: Grafton Books.

Browne, Ray Broadus, and Pat Browne, eds. 2001. *Guide to United States Popular Culture*. Madison: University of Wisconsin Press, 2001.

Burtt, Ben. 1993. "Ben Burtt—Sound Designer of Star Wars." Excerpt from *Star Wars Trilogy: The Definitive Collection*. Laserdisc. Republished in *Film Sound.org*. http://www.filmsound.org/starwars/burtt-interview.htm#Lightsabers.

Cage, John. 2013. *Silence: Lectures and Writings*. 50th Anniversary Edition. Middletown, CT: Wesleyan University Press.

Cazeaux, Clive. 2005. "Phenomenology and Radio Drama." *British Journal of Aesthetics* 45, no. 2: 157–174.

Chion, Michel. 1994. *Audio-Vision: Sound on Screen*. New York: Columbia University Press.

Choe, C. S., R. B. Welch, R. M. Gilford, and J. F. Juola. 1975. "The 'Ventriloquist Effect': Visual Dominance or Response Bias?" *Perception & Psychophysics* 18, no. 1: 55–60.

Coimbra, Fernando A. 2016. "Neolithic Art, Archaeoacoustics and Neuroscience." *Proceedings of Archaeoacoustics*. Myakka City, FL: The OTS Foundation.

Collins, Karen. 2002. "The Future Is Happening Already: Industrial Music, Dystopia and the Aesthetic of the Machine." PhD diss., University of Liverpool.

Collins, Karen. 2013. "Sonic Subjectivity and Auditory Perspective in *Ratatouille*." *Animation Journal* 8, no. 3: 283–299.

Collins, Karen. 2016. *The Beep Book: Documenting the History of Game Sound*, volumes 1 and 2. Waterloo: Ehtonal.

Collins, Karen, and Ruth Dockwray. 2015. "Sonic Proxemics and the Art of Persuasion: An Analytical Framework." *Leonardo Music Journal* 25: 53–56.

Collins, Karen, and Philip Tagg. 2001. "The Sonic Aesthetics of the Industrial: Re-Constructing Yesterday's Soundscape for Today's Alienation and Tomorrow's Dystopia." In *Sound Practice*, ed. John Drever, 101–108. Devon: UK/Ireland Soundscape Community.

Cone, E. 1968. *Musical Form and Musical Performance*. New York: Norton.

Connolly, Brian. 2015. "The Inner Ear as Musical Instrument." Paper presented at the Acoustical Society of America, Jacksonville, TN, November 6. https://acoustics.org/5amu1-the-inner-ear-as-a-musical-instrument-brian-connolly/.

Cox, Arnie. 2001. "The Mimetic Hypothesis and Embodied Musical Meaning." *Musicae Scientiae* 5, no. 2: 195–212.

Cox, Trevor J. 2008a. "The Effect of Visual Stimuli on the Horribleness of Awful Sounds." *Applied Acoustics* 69: 691–793.

Cox, Trevor J. 2008b. "Scraping Sounds and Disgusting Noises." *Applied Acoustics* 69: 1195–1204.

Cox, Trevor J. 2014. *The Sound Book*. New York: Norton.

Dixon, Michael J., Kevin A. Harrigan, Diane Santesso, C. Graydon, Jonathan A. Fugelsang, and Karen Collins. 2010. "The Impact of Sound in Modern Multiline Video Slot Machine Play." *Journal of Gambling Studies* 30, no. 4: 913–929.

Donaldson, Melissa. 2017. "Plutchik's Wheel of Emotions." SixSeconds, https://www.6seconds.org/2017/04/27/plutchiks-model-of-emotions/.

Dorritie, Frank. 2003. *The Handbook of Field Recording*. Vallejo, CA: Pro Audio Press.

Doyle, Peter. 2004. "From 'My Blue Heaven' to 'Race with the Devil': Echo, Reverb and (Dis)ordered Space in Early Popular Music Recording." *Popular Music* 23, no. 1: 31–49.

Dusan, Sorin V., Aram Lindahl, and Esge B. Andersen. 2013. "System and Method of Mixing Accelerometer and Microphone Signals to Improve Voice Quality in a Mobile Device." US Patent 9363596.

Eco, Umberto. 1979. *The Role of the Reader*. Bloomington: Indiana University Press.

Eddy, Cheryl. 2015. "The Surprising Objects Used to Make Gruesome Sounds on *The Walking Dead*." *Gizmodo,* April 2. https://io9.gizmodo.com/the-surprising-objects-used-to-make-gruesome-sounds-on-1695253108.

Ekman, Inger, and Michal Rinott. 2010. "Using Vocal Sketching for Designing Sonic Interactions." In *Proceedings of the 8th ACM Conference on Designing Interactive Systems, 2010*, 123–131. New York: ACM.

Evans, Fiona. "The Startling Truths of Old World Sparrows" *BBC's The Wire*, 2013. https://www.bbc.co.uk/writersroom/scripts/the-startling-truth-of-old-world-sparrows.

Ferguson, Stewart, and Sherry D. Ferguson. 1978. "Proxemics and Television: The Politician's Dilemma." *Canadian Journal of Communication* 4, no. 4: 26–35.

Fletcher, H., and W. A. Munson. 1933. "Loudness, Its Definition, Measurement and Calculation." *Journal of the Acoustical Society of America* 5: 82–108.

Freed, D. J. 1990. "Auditory Correlates of Perceived Mallet Hardness for a Set of Recorded Percussive Sound Events." *Journal of the Acoustical Society of America* 87: 311–322.

Fryer, L., J. Freeman, and L. Pring. 2014. "Touching Words Is Not Enough: How Visual Experience Influences Haptic–Auditory Associations in the 'Bouba–Kiki' Effect." *Cognition* 132, no. 2: 164–173.

Gaver, W. W. 1993. "What in the World Do We Hear? An Ecological Approach to Auditory Event Perception." *Ecological Psychology* 5: 1–29.

Gibson, David. 2018. *The Art of Mixing: A Visual Guide to Recording* New York: Routledge.

Gruters, Kurtis G., David L. K. Murphy, Cole D. Jenson, David W. Smith, Christopher A. Shera, and Jennifer M. Groh. 2018. "The Eardrums Move When the Eyes Move." *Proceedings of the National Academy of Sciences* 115, no. 6: E1309–E1318.

Hall, Edmond. 1963. "A System of Notation of Proxemic Behavior." *American Anthropologist* 41: 1003–1026.

Heidegger, Martin. 1962. *Being and Time*. New York: Harper & Row.

Holman, Tomlinson. 2002. *Sound for Film and Television*. Boston: Focal Press.

Horn, John. 2016. "*Mad Max* Sound Designer Mark Mangini Was Inspired by Moby Dick." *The Frame*. https://www.scpr.org/programs/the-frame/2016/02/02/46162/mad-max-sound-designer-mark-mangini-was-inspired-b/.

Hulusić, V., K. Debattista, V. Aggarwal, and A. Chalmers. 2010. "Maintaining Frame Rate Perception in Interactive Environments by Exploiting Audio-Visual Cross-Modal Interaction." *Visual Computer* 27, no. 1: 57–66.

Hunt, Elle. 2019. "Where There's a Horse, There's a Neigh: Why Must We Hear Animals on Screen?" *Guardian*, January 4. https://www.theguardian.com/tv-and-radio/shortcuts/2019/jan/04/where-theres-a-horse-theres-a-neigh-why-must-we-hear-animals-on-screen.

Ihde, Don. 1976. *Listening and Voice: A Phenomenology of Sound*. Athens: Ohio University Press.

Isaza, Miguel. 2009. "Walter Murch Special: The Concept of Worldizing." *Designing Sound*, October 7. http://designingsound.org/2009/10/07/walter-murch-special-the-concept-of-worldizing/.

Iversen, John R., Aniruddh D. Patel, and Kengo Ohgushi. 2008. "Perception of Rhythmic Grouping Depends on Auditory Experience." *Journal of the Acoustical Society of America* 124: 2263–2271.

Jarrett, Michael, and Walter Murch. 2000. "Sound Doctrine: An Interview with Walter Murch." *Film Quarterly* 53, no. 3: 2–11.

Jasen, Paul. 2016. *Low End Theory: Bass, Bodies and the Materiality of Sonic Experience*. New York: Bloomsbury.

Kincaid, Chris. 2016. "Manga Sound Effect Guide." *Japan Powered*, February 7. https://www.japanpowered.com/anime-articles/manga-sound-effect-guide.

Kitazaki, S. and Griffin, M. J. 1998. "Resonance Behaviour of the Seated Human Body and Effects of Posture." *Journal of Biomechanics* 31, no. 2: 143–149.

Köhler, Wolfgang. 1947. *Gestalt Psychology*. New York: Liveright Publishing.

Lane, J. D., S. J. Kasian, J. E. Owens, and G. R. Marsh. 1998. "Binaural Auditory Beats Affect Vigilance Performance and Mood." *Physiological Behaviour* 63, no. 2: 249–252.

Larkin, Edward. 1971. "Beethoven's Illness: A Likely Diagnosis." *Proceedings of the Royal Society of Medicine* 64: 493–496.

Lederman, S. J. 1979. "Auditory Texture Perception." *Perception* 8: 93–103.

Leman, Mark. 2008. *Embodied Music Cognition and Mediation Technology*. Cambridge, MA: MIT Press.

Maynes, Charles. 2004. "Worldizing: Take Studio Recordings into the Field to Make Them Sound Organic." *Editors Guild Magazine* 25, no. 2. http://magicshirt.com/Worldizing%20-%20Take%20Studio%20Recordings%20into%20the%20Field%20to%20Make%20them%20Sound%20Organic.pdf.

McCaffery, Steve, and bpNichol, eds. 1978. *Sound Poetry: A Catalogue*. Toronto: Underwich Editions.

McGurk, H., and J. MacDonald. 1976. "Hearing Lips and Seeing Voices." *Nature* 264, no. 5599: 746–748.

Mead, A. 2003. "Bodily Hearing: Physiological Metaphors and Musical Understanding." *Journal of Music Therapy* 43, no. 1: 1–19.

Meijerman, Lynn, Andrew Thean, and George Maat. 2005. "Earprints in Forensic Investigations." *Forensic Science, Medicine and Pathology* 1, no. 4: 247–256.

Mercedes. 2007. "Mercedes Introduces New Sound Logo to Compliment Updated Brand Design." October 25. http://www.emercedesbenz.com/Oct07/25_Mercedes_Introduces_New_Sound_Logo_To_Compliment_Updated_Brand_Design.html.

Merleau-Ponty, Maurice. 1998. *Phenomenology of Perception*. London: Routledge.

Moore, Allan. 1992. *Rock: The Primary Text: Developing a Musicology of Rock*. Aldershot: Ashgate.

Murch, Walter. 2000. "Stretching Sound to Help the Mind See." *New York Times*, October 1. http://www.filmsound.org/murch/stretching.htm.

Murch, Walter. 2005. "Dense Clarity—Clear Density." *Transom Review* 5. https://transom.org/2005/walter-murch/.

Niedenthal, P. M. 2007. "Embodying Emotion." *Science* 316: 1002–1005.

NPR. 2012a. "The Ballad of All Things Tearful." *All Things Considered*. https://wamu.org/story/12/02/13/the_ballad_of_the_tearful_why_some_songs_make_you_cry/.

NPR. 2012b. "Conquering Reverb: Behind Recorded Music's Oldest Sound Effect." July 6. https://www.npr.org/templates/transcript/transcript.php?storyId=156395020.

Nudds, Matthew, and Casey O'Callaghan. 2009. *Sounds and Perception: New Philosophical Essays*. Oxford: Oxford University Press.

Oakland Toy Lab. n.d. "Bite-Sized Boombox." *Instructables Workshop*. https://www.instructables.com/id/BoomBiteBox-listening-with-your-teeth/.

Oliveros, Pauline. 2005. *Deep Listening: A Composer's Guide to Sound Practice*. New York: iUniverse.

Ondaatje, Michael. 2002. *The Conversations: Walter Murch and the Art of Editing Film*. London: Bloomsbury.

Perlman, Marc. 2004. "Golden Ears and Meter Readers: The Contest for Epistemic Authority in Audiophilia." *Social Studies of Science* 34, no. 5: 783–807.

Petrosky, Max. n.d. "Frank Serafine Interview." *Tron Wiki*. https://tron.fandom.com/wiki/Tron_Wiki:Frank_Serafine_Interview.

Plutchik, Robert. 1991. *The Emotions*. New York: University Press of America.

Schafer, R. Murray. 1992. *A Sound Education: 100 Exercises in Listening and Soundmaking*. Tokyo: Shunjusha.

Schafer, R. Murray. 2009. "I Have Never Seen a Sound." *Journal of the Canadian Acoustical Association*, 37 no. 3. https://jcaa.caa-aca.ca/index.php/jcaa/article/view/2123.

Sekuler, R., A. B. Sekuler, and R. Lau. 1997. "Sound Alters Visual Motion Perception." *Nature* 385, no. 6614: 308.

Shaha, Alom. n.d. "Wave Machine." National STEM Learning Centre. https://www.stem.org.uk/resources/elibrary/resource/27031/wave-machine#&gid=undefined&pid=1.

Shams, L., Y. Kamitani, and S. Shimojo. 2000. "Illusions: What You See Is What You Hear." *Nature* 408, no. 6814: 788.

Shiga, David. 2000. "Sound Can Leap across a Vacuum after All." *New Scientist*, September 29. https://www.newscientist.com/article/mg20827804-600-sound-can-leap-across-a-vacuum-after -all/.

Sigismondi, Gino, Tim Vear, and Rick Waller. 2014. "Shure Microphone Techniques for Record-ing." http://cdn.shure.com/publication/upload/837/microphone_techniques_for_recording_ english.pdf.

Spence, C., and M. U. Shanker. 2010. "The Influence of Auditory Cues on the Perception of, and Responses to, Food and Drink." *Journal of Sensory Studies* 25, no. 3: 406–430.

Sterne, Jonathan. 2006a. "The Death and Life of Digital Audio." *Interdisciplinary Science Reviews* 31, no. 4: 339–348.

Sterne, Jonathan. 2006b. "The MP3 as Cultural Artifact." *New Media & Society* 8, no. 5: 825–842.

Tagg, Philip. 2003. *Kojak: 50 Seconds of Television Music*. New York: Mass Media Music Scholars' Press.

Tajadura-Jiménez Ana, Aleksander Väljamäe, Erkin Asutay, and Daniel Västfjäll. 2010. "Embodied Auditory Perception: The Emotional Impact of Approaching and Receding Sound Sources." *Emotion* 10, no. 2: 216–229.

Tompkins, Dave. 2010. *How to Wreck a Nice Beach: The Vocoder from World War II to Hip-Hop, The Machine Speaks*. Chicago: Stop Smiling Books.

Van Eck, Cathy. 2017. *Between Air and Electricity: Microphones and Loudspeakers as Musical Instruments*. New York: Bloomsbury.

Verma, Neil. 2012. *Theatre of the Mind: Imagination, Aesthetics and American Radio Drama*. Chicago: University of Chicago Press.

Weinel, Jonathan. 2018. *Inner Sound: Altered States of Consciousness in Electronic Music and Audio-Visual Media*. Oxford: Oxford University Press.

Westerkamp, Hildegard. 2007. "Soundwalking." In *Autumn Leaves, Sound and the Environment*, ed. Angus Carlyle. Paris: Double Entre.

White, Charlie. 2007. "Pear Cable Chickens Out of $1,000,000 Challenge, We Search for Answers." *Gizmodo*, October 26. https://gizmodo.com/pear-cable-chickens-out-of-1-000-000 -challenge-we-sea-315250.

Whitwell, Tom. 2005. "Tiny Music Makers Pt 3: The THX Sound." *Music Thing*. http:// musicthing.blogspot.com/2005/05/tiny-music-makers-pt-3-thx-sound.html.

Wildes, R. P., and W. A. Richards. 1988. "Recovering Material Properties from Sound." In *Natural Computation*, ed. W. A. Richards, 356–363. Cambridge, MA: MIT Press.

Wright, Benjamin. 2015. "Atmos Now: Dolby Laboratories, Mixing Ideology and Hollywood Sound Production." In *Living Stereo: History, Culture, Multichannel Sound*, ed. Paul Théberge, Kyle Devine, and Tom Everett. New York: Bloomsbury.

Yost, William A. 2018. "Auditory Motion Parallax." *Proceedings of the National Academy of Sciences of the United States of America* 115, no. 16: 3998–4000.

Zwisler, Evan. 2017. "Tales from the Toilet: Truth vs. Fiction in Musical Stories of the Bathroom." *Flypaper*, November 17. https://flypaper.soundfly.com/play/tales-toilet-truth-fiction-musical-stories -of-the-bathroom/.

Audiovisual References

Apocalypse Now. American Zoetrope, 1979.

Audio Defence (Zombie Arena). Somethin' Else, 2014.

Can. *Flow Motion*. Harvest Records, 1976.

Cher. *Believe*. Warner Bros, 1998.

Gerald McBoing-Boing. UPA Pictures, 1956.

Hellblade. Ninja Theory, 2017.

Jaws. Universal, 1975.

Papa Sangre. Somethin' Else, 2010.

Pearl Jam. *Binaural*. Epic, 2000.

Psycho. Paramount, 1960.

Quiet Place, A. Paramount, 2018.

Rain People, The. American Zoetrope, 1969.

Sniper Elite 4. Rebellion, 2017.

Stalker. Mosfilm, 1979.

Star Wars. LucasFilm, 1976.

Tron. Walt Disney, 1982.

Walking Dead, The. AMC, 2010–.

Wall-E. Pixar, 2008.

Index